Lecture Notes in Control and Information Sciences

Edited by M. Thoma and A. Wyner

For information about Vols. 1–116 please contact your bookseller or Springer-Verlag

Lecture Notes
in Control and Information Sciences   179

Editors: M. Thoma and W. Wyner

Lecture Notes
in Control and Information Sciences   179

Editors: M. Thoma and W. Wyner

Z.H. Jiang, W. Schaufelberger

Block Pulse Functions and Their Applications in Control Systems

Springer-Verlag
Berlin Heidelberg GmbH

Advisory Board

Authors

Zhihua Jiang
Walter Schaufelberger
Projekt-Zentrum IDA
ETH-Zentrum
8092 Zürich
Switzerland

ISBN 978-3-540-55369-4   ISBN 978-3-540-47046-5 (eBook)
DOI 10.1007/978-3-540-47046-5

Typesetting: Camera ready by authors

60/3020 5 4 3 2 1 0 Printed on acid-free paper

Preface

Block pulse functions have been studied and applied extensively in the past fifteen years as a basic set of functions for signal characterizations in systems science and control. All these studies and applications show that block pulse functions may have definite advantages for problems involving integrals and derivatives due to their clearness in expressions and their simplicity in formulations. After original problems are transformed into their corresponding algebraic expressions, piecewise constant approximate solutions can be computed efficiently to show the tendency of the exact solutions under flexible weighting of the accuracy of the results and the size of computations.

In such a way, block pulse functions may be successfully used in many basic problems in the area of systems and control, such as

- transforms
- simulations
- system analysis
- system synthesis
- system identification

and so on.

Our purpose of writing this book is to present the principles and techniques of block pulse functions systematically and uniformly, so that the readers who are interested in using block pulse functions in their own problems can obtain an overview of the existing methods and the developing tendency of this area. Although our discussions are restricted to block pulse functions, the ideas we use here are general and can be used as a reference to other orthogonal functions and orthogonal polynomials, e.g. Walsh functions, Chebyshev polynomials, Legendre polynomials, Laguerre polynomials and so on, because the manipulation of problems based on these orthogonal functions or polynomials are very similar to the ones of the block pulse functions.

In comparison with other basis functions or polynomials, the block pulse functions can lead more easily to recursive computations to solve concrete problems. Therefore various recursive formulas are discussed with emphasis in this book e.g. the formulas of multiple integrals and of block pulse difference equations for various dynamic systems under the input-output representation and state space representation.

The book has been written for staff, students and engineers in industry with interest in digital signal processing for system analysis and design. In our view, the block pulse function techniques are very well suited for rapid prototyping of software solutions in these areas.

Zurich, Novenber 1991

Zhihua Jiang

Walter Schaufelberger

Contents

Nomenclature

Following is a list of principal symbols which appear frequently in the expressions of block pulse functions in this book. For convenience, each symbol is followed by a brief description and the section number where it is first used or formally defined.

$\phi_i(t)$ block pulse function   (1.1)

$\varphi_i(t)$ orthonormal block pulse function   (1.2)

$\phi_\lambda(t),\ \phi_\lambda^*(t)$ auxiliary function with a single block pulse   (2.2)

$[0, T)$ definition interval of block pulse functions   (1.1)

h width of block pulses   (1.1)

m number of total block pulse functions   (1.1)

$\bar{f}_i$ discrete value of $f(t)$ at $t = ih$ (1.3)

f_i block pulse coefficient of $f(t)$ (1.2)

$\hat{f}_m(t)$ block pulse series of $f(t)$ (1.2)

$f^{(i)}(t)$ ith order derivative of $f(t)$ (2.2)

$f_0^{(i)}$ initial value of $f^{(i)}(t)$ (2.2)

$\Phi(t)$ block pulse function vector   (2.1)

F block pulse coefficient vector of $f(t)$ (2.1)

E constant vector with all entries ones   (2.1)

I unit matrix   (2.1)

Δ_i vector with its ith entry one and remaining entries zeros   (2.1)

D_F diagonal matrix related to vector F (2.1)

H one step delay operational matrix   (2.1)

$\overrightarrow{H}_q$ matrix used in block pulse series of function $f(t - \tau)$ (2.1)

$\overleftarrow{H}_q$ matrix used in block pulse series of function $f(\tau - t)$ (2.1)

J_F convolution operational matrix. (2.2)

$\tilde{J}_F$ another convolution operational matrix (2.3)

P conventional integration operational matrix (2.2)

$\bar{P}$ improved integration operational matrix (5.1)

P_k generalized integration operational matrix (5.2)

$P_{i,j}$ extended integration operational matrix (5.5)

$p_{k,j}$ part of the entry of matrix P_k (5.2)

$p_{i,j,k,l}$ part of the entry of matrix $P_{i,j}$ (5.5)

$G^{(t)}$ block pulse transfer matrix (6.1)

$g_i^{(t)}, g_{i,j}^{(t)}$ entry of block pulse transfer matrix (6.2)

B block pulse operator (3.1)

$F^s(z)$ block pulse transform of $F(s)$ (4.1)

A_i parameter in block pulse difference equation (7.1)

a_i parameter in block pulse regression equation (7.1)

Chapter 1

Introduction

Block pulse functions are a set of orthogonal functions with piecewise constant values and are usually applied as a useful tool in the analysis, synthesis, identification and other problems of control and systems science. This set of functions was first introduced to electrical engineers by Harmuth in 1969, but for about seven years it has not received any attention with regard to practical applications. Until the middle of the seventies, several researchers (Gopalsami and Deekshatulu, 1976a, 1976b; Chen, Tsay and Wu, 1977; Sannuti, 1977) discussed the block pulse functions and their operational matrix for integration in order to reduce the complexity of expressions in solving certain control problems via Walsh functions. Since then, the block pulse functions have been extensively applied due to their simple and easy operations. In this chapter, we will introduce the basic concepts of block pulse functions.

1.1 Definition of block pulse functions

As proposed by many authors, a set of block pulse functions $\phi_i(\lambda)$ $(i = 1, 2, \ldots, m)$ is usually defined in the unit interval $\lambda \in [0, 1)$ as:

$$\phi_i(\lambda) = \begin{cases} 1 & \text{for } (i-1)/m \le \lambda < i/m \\ 0 & \text{otherwise} \end{cases} \tag{1.1}$$

Equation (1.1) shows that the unit interval $\lambda \in [0, 1)$ is divided into m equidistant subintervals, and the ith block pulse function $\phi_i(\lambda)$ has only one rectangular pulse of unit height in the ith subinterval $\lambda \in [(i-1)/m, i/m)$.

In fact, the restriction of an interval with unit length is not secessary in the definition of block pulse functions. If we use the substitution $t = T\lambda$ in (1.1), we can obtain a similar definition of block pulse functions with regard to an interval of arbitrary length to meet the requirements of various problems. Therefore we define in this book the set of block pulse functions $\phi_i(t)$ $(i = 1, 2, \ldots, m)$ in the interval $t \in [0, T)$:

$$\phi_i(t) = \begin{cases} 1 & \text{for } (i-1)T/m \le t < iT/m \\ 0 & \text{otherwise} \end{cases} \tag{1.2}$$

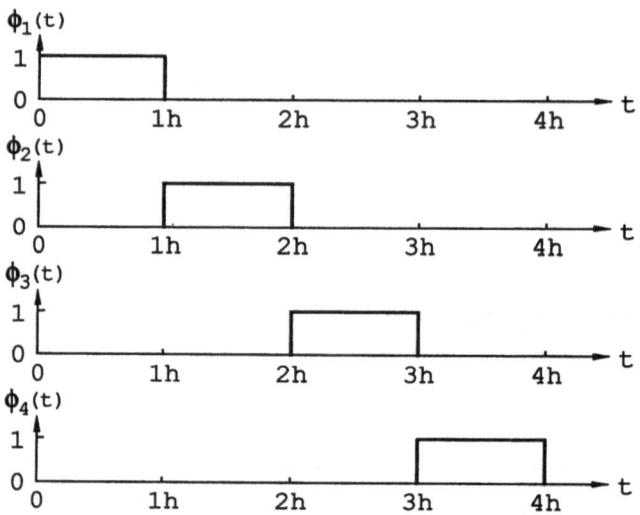

Figure 1.1: Block pulse functions, $(m = 4)$.

where m is an arbitrary positive integer. As an example, a set of block pulse functions with $m = 4$ is illustrated in Figure 1.1.

If problems are studied in an interval $\tau \in [a, b)$, we can always transform them into the equivalent problems in the interval $t \in [0, T)$ through:

$$t = \frac{\tau - a}{b - a} T \qquad (1.3)$$

If problems are studied in an infinite interval $t \in [0, \infty)$, we can always choose sufficiently large values of T to meet the requirements of the interval length. After the interval length T is determined, the accuracy of the results obtained from the block pulse function technique can be adjusted properly by the integer m, or expressing more clearly, by T/m. For the convenience of expressions, we usually denote the equidistant subinterval or the width of block pulses as $h = T/m$ in this book.

1.2   Elementary properties of block pulse functions

In solving certain problems of control and systems science, the advantages of using the block pulse function technique are their easy operations and satisfactory approximations. These advantages are due to the distinct properties of block pulse functions. The elementary properties are as follows.

1. Disjointness. The block pulse functions are disjoined with each other in the interval

$t \in [0, T)$:

$$\phi_i(t)\phi_j(t) = \begin{cases} \phi_i(t) & \text{for } i = j \\ 0 & \text{for } i \neq j \end{cases} \qquad (1.4)$$

where $i, j = 1, 2, \ldots, m$. This property can directly be obtained from the definition of block pulse functions.

Owing to the disjointness of block pulse functions, the joint terms will disappear in each subinterval when multiplication, division and some other operations are applied based on the block pulse series expansion of functions, as we will discuss in the next several chapters. This property will simplify greatly the calculations in the problems.

2. Orthogonality. The block pulse functions are orthogonal with each other in the interval $t \in [0, T)$:

$$\int_0^T \phi_i(t)\phi_j(t)dt = \begin{cases} h & \text{for } i = j \\ 0 & \text{for } i \neq j \end{cases} \qquad (1.5)$$

where $i, j = 1, 2, \ldots, m$. This property can directly be obtained from the disjointness of block pulse functions.

The orthogonal property of block pulse functions is the basis of expanding functions into their block pulse series. An arbitrary real bounded function $f(t)$, which is square integrable in the interval $t \in [0, T)$, can be expanded into a block pulse series in the sense of minimizing the mean square error between $f(t)$ and its approximation:

$$\begin{aligned} f(t) &\doteq \hat{f}_m(t) \\ &= \sum_{i=1}^{m} f_i \phi_i(t) \end{aligned} \qquad (1.6)$$

where $\hat{f}_m(t)$ is the block pulse series of the original function $f(t)$, and f_i is the block pulse coefficient with respect to the ith block pulse function $\phi_i(t)$.

3. Completeness. The block pulse function set is complete when m approaches infinity. This means that we have (Tolstov, 1962):

$$\int_0^T f^2(t)dt = \sum_{i=1}^{\infty} f_i^2 \|\phi_i(t)\|^2 \qquad (1.7)$$

for any real bounded function $f(t)$ which is square integrable in the interval $t \in [0, T)$. Here, the expression:

$$\|\phi_i(t)\| = \left(\int_0^T \phi_i^2(t)dt \right)^{\frac{1}{2}} \qquad (1.8)$$

is the norm of $\phi_i(t)$.

When block pulse functions are used to solve problems, original functions are usually approximated by their block pulse series. Therefore it is important to know, whether the mean square error between a given function and its approximate block pulse series can become arbitrary small under certain conditions. The completeness of block pulse functions guarantees that an arbitrarily small mean square error can be obtained for a real bounded function, which has only a finite number of discontinuous points in the interval $t \in [0, T)$, by increasing the number of terms in the block pulse series. The details of this completeness will be discussed in Section 1.4.

Equation (1.8) indicates that the block pulse functions defined in (1.2) are not orthonormal. The corresponding orthonormal block pulse function set can easily be generated from (1.2) as:

$$\varphi_i(t) = \frac{1}{\sqrt{h}} \phi_i(t) \tag{1.9}$$

or directly defined as:

$$\varphi_i(t) = \begin{cases} \dfrac{1}{\sqrt{h}} & \text{for } (i-1)h \le t < ih \\ 0 & \text{otherwise} \end{cases} \tag{1.10}$$

But usually, we prefer to use the definition (1.2) to comply with the expressions in the general literature.

1.3 Block pulse series expansions

The block pulse series expansion in (1.6) indicates that the originl function $f(t)$ can be approximated by a piecewise constant function $\hat{f}_m(t)$. The criterion of this approximation is that the mean square error between $f(t)$ and $\hat{f}_m(t)$ in the interval $t \in [0, T)$:

$$\varepsilon = \frac{1}{T} \int_0^T \left(f(t) - \sum_{j=1}^m f_j \phi_j(t) \right)^2 dt \tag{1.11}$$

reaches its minimum. Based on this criterion, we can determine the values of the block pulse coefficients.

From

$$\frac{\partial \varepsilon}{\partial f_i} = -\frac{2}{T} \int_0^T \left(f(t) - \sum_{j=1}^m f_j \phi_j(t) \right) \phi_i(t) dt = 0 \tag{1.12}$$

and the orthogonal property of block pulse functions, we obtain:

$$\int_{(i-1)h}^{ih} f_i \phi_i(t) dt = \int_{(i-1)h}^{ih} f(t) \phi_i(t) dt \tag{1.13}$$

Therefore the necessary condition of minimizing the mean square error ε yields the block pulse coefficient f_i ($i = 1, 2, \ldots, m$):

$$\begin{aligned} f_i &= \frac{1}{h} \int_0^T f(t) \phi_i(t) dt \\ &= \frac{1}{h} \int f(t) dt \Big|_{(i-1)h}^{ih} \end{aligned} \tag{1.14}$$

Moreover, since the expression:

$$\frac{\partial^2 \varepsilon}{\partial f_i \partial f_j} = \begin{cases} \dfrac{2}{m} > 0 & \text{for } i = j \\[2mm] 0 & \text{for } i \neq j \end{cases} \tag{1.15}$$

holds for $i, j = 1, 2, \ldots, m$, it is easy to verify the determinants for $k = 1, 2, \ldots, m$:

$$D_k = \begin{vmatrix} \dfrac{\partial^2 \varepsilon}{\partial f_1^2} & \dfrac{\partial^2 \varepsilon}{\partial f_1 \partial f_2} & \cdots & \dfrac{\partial^2 \varepsilon}{\partial f_1 \partial f_k} \\[3mm] \dfrac{\partial^2 \varepsilon}{\partial f_2 \partial f_1} & \dfrac{\partial^2 \varepsilon}{\partial f_2^2} & \cdots & \dfrac{\partial^2 \varepsilon}{\partial f_2 \partial f_k} \\[3mm] \vdots & \vdots & \ddots & \vdots \\[3mm] \dfrac{\partial^2 \varepsilon}{\partial f_k \partial f_1} & \dfrac{\partial^2 \varepsilon}{\partial f_k \partial f_2} & \cdots & \dfrac{\partial^2 \varepsilon}{\partial f_k^2} \end{vmatrix} > 0 \tag{1.16}$$

This is just the sufficient condition for the minimum of the mean square error ε under the block pulse coefficients f_i ($i = 1, 2, \ldots, m$) which are determined by (1.14).

Equation (1.14) reveals the geometrical meaning of the block pulse coefficients. In fact, the ith block pulse coefficient f_i is the integral mean value of the original function $f(t)$ in the ith subinterval. In other words, the ith block pulse coefficient f_i is related to the area in the subinterval $t \in [(i-1)h, ih)$ under the curve of $f(t)$. From this point of view, we can easily expand the function $f(t)$ into its block pulse series, regardless whether the function is given analytically by formula, or graphically by curve. In practice, if the width of block pulses h is small enough, the block pulse coefficients can also be determined approximately by the simple relation:

$$f_i \doteq \frac{1}{2}(f_{i-1} + f_i) \tag{1.17}$$

where f_{i-1} and f_i are the values of $f(t)$ at the time instants $t = (i-1)h$ and $t = ih$, respectively. Equation (1.17) reduces computations and shows that each block pulse coefficient is approximated by the mean value of the original function at two end points of the corresponding subinterval. In other words, the area bounded by the original function in each subinterval is now approximated by the corresponding trapezoid. Here is a simple example to show both these block pulse series expansions.

Example 1.1 Expand $f(t) = t^2$ into its block pulse series in the interval $t \in [0, 1)$ with $m = 4$, using both the exact and approximate formulas (1.14) and (1.17).

Since the exact formula (1.14) gives the block pulse coefficients:

$$\begin{aligned} f_1 &= 4 \int_0^1 t^2 \phi_1(t) dt = \frac{1}{48} \\[2mm] f_2 &= 4 \int_0^1 t^2 \phi_2(t) dt = \frac{7}{48} \\[2mm] f_3 &= 4 \int_0^1 t^2 \phi_3(t) dt = \frac{19}{48} \\[2mm] f_4 &= 4 \int_0^1 t^2 \phi_4(t) dt = \frac{37}{48} \end{aligned}$$

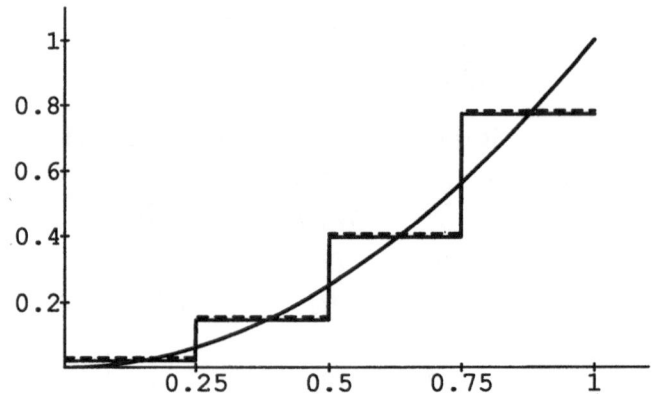

— : from equation (1.14),   - - - : from equation (1.17)

Figure 1.2: Block pulse series of $f(t) = t^2$, $(m = 4)$.

the block pulse series of $f(t) = t^2$ is:

$$f(t) \doteq \frac{1}{48}\phi_1(t) + \frac{7}{48}\phi_2(t) + \frac{19}{48}\phi_3(t) + \frac{37}{48}\phi_4(t)$$

Whereas the block pulse coefficients are:

$$
\begin{aligned}
f_1 &= \frac{1}{2}\left(0^2 + \left(\frac{1}{4}\right)^2\right) = \frac{1}{32} \\
f_2 &= \frac{1}{2}\left(\left(\frac{1}{4}\right)^2 + \left(\frac{1}{2}\right)^2\right) = \frac{5}{32} \\
f_3 &= \frac{1}{2}\left(\left(\frac{1}{2}\right)^2 + \left(\frac{3}{4}\right)^2\right) = \frac{13}{32} \\
f_4 &= \frac{1}{2}\left(\left(\frac{3}{4}\right)^2 + 1^2\right) = \frac{25}{32}
\end{aligned}
$$

according to the approximate formula (1.17). Both these results are illustrated in Figure 1.2. We can notice that a better block pulse series approximation can be obtained if a smaller width of block pulses is chosen. But in such cases, more block pulse coefficients are involved in the computations. Therefore in choosing a proper integer m for block pulse functions, a compromise should be made between the accuracy of approximations and the size of computations.

1.4   Convergence of block pulse series

In approximating original functions via block pulse series, a natural question will arise, i.e. is the block pulse function set complete? If the function set is incomplete, we cannot guarantee that an arbitrary small mean square error can be obtained for any given functions by increasing the number of terms in the series. In fact, the answer of this question is positive. It was answered by Kwong and Chen (1981a) through the discussion about the convergence of block pulse series.

We first assume that the original function $f(t)$ is a real bounded continuous function in the interval $t \in [0, T)$. This continuous function can be expanded into its block pulse series according to (1.14). But now we write (1.14) in a more clear form:

$$f_i = \frac{m}{T} \int_{(i-1)T/m}^{iT/m} f(t)dt \tag{1.18}$$

First, we study the first block pulse coefficient:

$$f_1 = \frac{m}{T} \int_0^{T/m} f(t)dt \tag{1.19}$$

If the interval $t \in [0, T/m)$ is partitioned into p subintervals of length Δt_k ($k = 1, 2, \ldots, p$) and if an arbitrary point ξ_k is chosen in each subinterval, we obtain the Riemann sum:

$$R(f) = \sum_{k=1}^{p} f(\xi_k)\Delta t_k \tag{1.20}$$

Let Δt_λ denote the maximum of Δt_k, then $R(f)$ has a limit as Δt_λ approaches zero because $f(t)$ is continuous in the interval $t \in [0, T)$. This limit is the Riemann integral of $f(t)$ in the subinterval $t \in [0, T/m)$:

$$
\begin{aligned}
\int_0^{T/m} f(t)dt &= \lim_{\Delta t_\lambda \to 0} R(f) \\
&= \lim_{\Delta t_\lambda \to 0} \sum_{k=1}^{p} f(\xi_k)\Delta t_k
\end{aligned}
\tag{1.21}
$$

Since the partition of Δt_k and the selection of ξ_k are arbitrary, we must obtain the same limit of $R(f)$ if we choose $\Delta t_k = \Delta t = T/(pm)$ and $\xi_k = (k-1)\Delta t$. Thus, from (1.19) and (1.21), the first block pulse coefficient becomes:

$$f_1 = \lim_{\Delta t \to 0} \sum_{k=1}^{p} \frac{1}{p} f((k-1)\Delta t) \tag{1.22}$$

Noticing the existence of the limits:

$$\lim_{m \to \infty} \Delta t = \lim_{m \to \infty} \frac{T}{pm} = 0 \tag{1.23}$$

and

$$\lim_{m \to \infty} (k-1)\Delta t = \lim_{m \to \infty} \frac{(k-1)T}{pm} = 0 \tag{1.24}$$

we obtain:

$$\lim_{m \to \infty} f_1 = f(0) \tag{1.25}$$

Equation (1.25) shows that if the function $f(t)$ is continuous in the interval $t \in [0, T)$, the first block pulse coefficient f_1 in the block pulse series takes a limiting value of $f(0)$ as m approaches infinity.

For extending the obtained result to the ith block pulse coefficient, we shift the function $f(t)$ in the negative direction by an arbitrary $t_i \in (0, T)$. Noticing that the integral after shifting:

$$\frac{m}{T} \int_0^{T/m} f(t + t_i) dt \tag{1.26}$$

is equivalent to the integral before shifting:

$$\frac{m}{T} \int_{t_i}^{t_i + T/m} f(t) dt = f_i \tag{1.27}$$

we can obtain:

$$\lim_{m \to \infty} f_i = f(t_i) \tag{1.28}$$

by using the same approach as described above. From (1.25) and (1.28), we can conclude that the block pulse series $\hat{f}_m(t)$ in (1.6) pointwise converges to the real bounded continuous function $f(t)$ as m approaches infinity. But in such cases, the number of block pulse coefficients also approaches infinity.

For a real bounded function $g(t)$ which is piecewise continuous in the interval $t \in [0, T)$, if we neglect a finite number of discontinuous points, the same procedure discussed above can also be applied to each continuous segment of $g(t)$. Therefore we can conclude that the block pulse series $\hat{g}_m(t)$ in (1.6) pointwise converges to $g(t)$ as m approaches infinity, except possibly at a finite number of discontinuous points. This is the convergence property of block pulse series.

Above we discussed the convergence of block pulse series using the well established results of the Riemann integral (Shilov and Gurevich, 1966). The same result can also be obtained from the delta sequence of positive type (Korevaar, 1968) and the theory of generalized functions (Lighthill, 1958). Details of this proof can be found in the paper of Kwong and Chen (1981a).

For testing the completeness of the block pulse function set, the general criterion (Tolstov, 1962) can be presented concretely as follows. If for every real bounded continuous function $f(t)$ in the interval $t \in [0, T)$ and for any number $\epsilon > 0$, there exists an integer m so that the block pulse series (1.6) can satisfy the inequality:

$$\int_0^T \left(f(t) - \hat{f}_m(t) \right)^2 dt \leq \epsilon \tag{1.29}$$

then the block pulse function set $\{\phi_i(t);\ i = 1, 2, \ldots, m\}$ is complete. Since we have already proved the pointwise convergence of block pulse series $\hat{f}_m(t)$ of the continuous function $f(t)$ as m approaches infinity, it follows immediately that (1.29) must hold. Therefore the block pulse function set is a complete set when m approaches infinity.

1.5   Two dimensional block pulse functions

In the previous sections, we discussed the block pulse functions which have a single variable. These discussions can also be extended to the functions which have two variables. To distinguish them, we usually call the block pulse functions containing one variable as one-dimensional (1D) block pulse functions and those containing two variables as two-dimensional (2D) block pulse functions. Since our discussions in this book are concentrated in one-dimensional block pulse functions, the details of two-dimensional block pulse functions will not be given here. Following are only some short discussions about the 2D block pulse functions as a complement of the later chapters.

Similar to the 1D case, a set of 2D block pulse functions $\phi_{i_1,i_2}(t_1, t_2)$ $(i_1 = 1, 2, \ldots, m_1;$ $i_2 = 1, 2, \ldots, m_2)$ is defined in the region of $t_1 \in [0, T_1)$ and $t_2 \in [0, T_2)$ as:

$$\phi_{i_1,i_2}(t_1, t_2) = \begin{cases} 1 & \text{for } (i_1 - 1)h_1 \leq t_1 < i_1 h_1 \\ & \quad \text{and } (i_2 - 1)h_2 \leq t_2 < i_2 h_2 \\ 0 & \text{otherwise} \end{cases} \tag{1.30}$$

where m_1, m_2 are arbitrary positive integers, and $h_1 = T_1/m_1$, $h_2 = T_2/m_2$.

Similar to the 1D case, the 2D block pulse functions are disjoined with each other:

$$\phi_{i_1,i_2}(t_1, t_2)\phi_{j_1,j_2}(t_1, t_2) = \begin{cases} \phi_{i_1,i_2}(t_1, t_2) & \text{for } i_1 = j_1 \text{ and } i_2 = j_2 \\ 0 & \text{otherwise} \end{cases} \tag{1.31}$$

and are orthogonal with each other:

$$\int_0^{T_1} \int_0^{T_2} \phi_{i_1,i_2}(t_1, t_2)\phi_{j_1,j_2}(t_1, t_2)dt_2 dt_1 = \begin{cases} h_1 h_2 & \text{for } i_1 = j_1 \text{ and } i_2 = j_2 \\ 0 & \text{otherwise} \end{cases} \tag{1.32}$$

in the region of $t_1 \in [0, T_1)$ and $t_2 \in [0, T_2)$, where $i_1, j_1 = 1, 2, \ldots, m_1$; $i_2, j_2 = 1, 2, \ldots,$ m_2. Also similar to the 1D case, the 2D block pulse function set is complete when both m_1 and m_2 approach infinity.

We can also expand a two-variable function $f(t_1, t_2)$ into its block pulse series:

$$f(t_1, t_2) \doteq \sum_{i_1=1}^{m_1} \sum_{i_2=1}^{m_2} f_{i_1,i_2} \phi_{i_1,i_2}(t_1, t_2) \tag{1.33}$$

through determining the block pulse coefficients:

$$f_{i_1,i_2} = \frac{1}{h_1 h_2} \int_{(i_1-1)h_1}^{i_1 h_1} \int_{(i_2-1)h_2}^{i_2 h_2} f(t_1, t_2)dt_2 dt_1 \tag{1.34}$$

such that the mean square error between $f(t_1, t_2)$ and its block pulse series $\hat{f}_{m1,m2}(t_1, t_2)$ in the region of $t_1 \in [0, T_1)$ and $t_2 \in [0, T_2)$ is minimal:

$$\varepsilon = \frac{1}{T_1 T_2} \int_0^{T_1} \int_0^{T_2} \left(f(t_1, t_2) - \sum_{j_1=1}^{m_1} \sum_{j_2=1}^{m_2} f_{j_1,j_2} \phi_{j_1,j_2}(t_1, t_2) \right)^2 dt_2 dt_1 \tag{1.35}$$

Since each two-dimensional block pulse function takes only one value in its subregion, the 2D block pulse functions can be expressed by the two 1D block pulse functions:

$$\phi_{i_1,i_2}(t_1, t_2) = \phi_{i_1}(t_1)\psi_{i_2}(t_2) \qquad (1.36)$$

where $\phi_{i_1}(t_1)$ and $\psi_{i_2}(t_2)$ are the one-dimensional block pulse functions related to the variables t_1 and t_2, respectively. From this relation, we notice that the two-dimensional block pulse functions can be manipulated in a similar way to the one-dimensional block pulse functions.

The above discussion can be extended similarly to multi-dimensional (MD) block pulse functions further. Usually, the two- and multi-dimensional block pulse functions can be applied to problems which are related to partial differential equations. Discussions about the 2D and MD block pulse functions and their applications can be found in the papers of Stavroulakis and Tzafestas (1980), Rao and Srinivasan (1980b), Shih and Hwang (1982), Hsu and Cheng (1982), Hwang, Guo and Shih (1983), Marszalek (1983), Nath and Lee (1983), Perng and Chen (1985), Chen and Hsu (1987), Ning and Jiong (1988).

Chapter 2

Operations of block pulse series

The block pulse function technique is frequently applied in control and systems science to reduce the complexity of numerical problems. The main idea of this technique is rather simple. As shown in Figure 2.1, the original functions in the problems are first expanded into their block pulse series, and then the operations of block pulse series are applied to obtain the piecewise constant approximate solutions. Since the operations of block pulse series are much simpler than those of the original functions, the goal of computational reduction can be attained. In this chapter, we will discuss the operation rules of block pulse series which are the basis of the block pulse function technique.

2.1 Vector forms of block pulse series

Continuous functions can be expanded into their block pulse series. In order to express these series in plain and compact forms, we denote the m block pulse functions

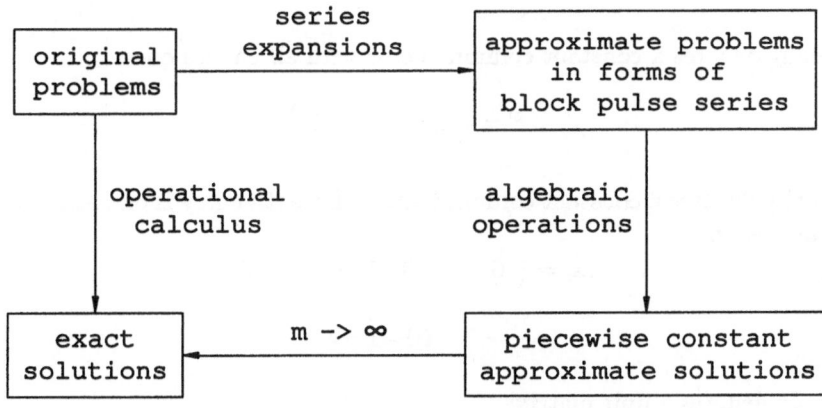

Figure 2.1: Basic idea of block pulse function technique.

together as a block pulse function vector:

$$\Phi(t) = \left(\begin{array}{cccc} \phi_1(t) & \phi_2(t) & \cdots & \phi_m(t) \end{array} \right)^T \tag{2.1}$$

and the m block pulse coefficients of the function $f(t)$ together as a block pulse coefficient vector of $f(t)$:

$$F = \left(\begin{array}{cccc} f_1 & f_2 & \cdots & f_m \end{array} \right)^T \tag{2.2}$$

where T denotes transpose. Using these notations, the formulas about the block pulse series expansion (1.6) and (1.14) are now:

$$\begin{aligned} f(t) &\doteq \hat{f}_m(t) \\ &= F^T \Phi(t) \\ &= \Phi^T(t) F \end{aligned} \tag{2.3}$$

and

$$F = \frac{1}{h} \int_0^T f(t)\Phi(t)dt \tag{2.4}$$

Extending this vector form expression to the two-dimensional block pulse functions, the block pulse series of $f(t_1, t_2)$ in (1.33) can be written as:

$$f(t_1, t_2) \doteq \Phi^T(t_1) F \Psi(t_2) \tag{2.5}$$

where $\Phi(t_1)$ and $\Psi(t_2)$ are respectively m_1- and m_2-dimensional block pulse function vectors, and F is an $m_1 \times m_2$ block pulse coefficient matrix.

For the convenience of expressions in the later discussions, we also introduce some notations for the commonly used vectors and matrices. Since these notations appear mainly in the expressions of block pulse series which have m terms, they usually stand for $m \times 1$ vectors and $m \times m$ matrices, unless special notes are given.

• The uppercase letters usually denote block pulse coefficient vectors if the corresponding lowercase letters express original functions, e.g. the vector F and the function $f(t)$ in (2.3).

• E usually denotes a constant column vector with all entries ones:

$$E = \left(\begin{array}{cccc} 1 & 1 & \cdots & 1 \end{array} \right)^T \tag{2.6}$$

• Δ_i usually denotes a constant column vector whose ith entry is one and the remaining entries are zeros:

$$\Delta_i = \left(\begin{array}{ccccccc} 0 & \cdots & 0 & 1 & 0 & \cdots & 0 \end{array} \right)^T \tag{2.7}$$
$$\uparrow$$
$$i\text{th-entry}$$

• I usually denotes a unit matrix:

$$I = \text{diag}(1, 1, \ldots, 1) \tag{2.8}$$

- D_C usually denotes a diagonal matrix whose diagonal entries are related to a constant vector C:

$$D_C = \text{diag}(c_1, c_2, \ldots, c_m) \tag{2.9}$$

where

$$C = \begin{pmatrix} c_1 & c_2 & \cdots & c_m \end{pmatrix}^T \tag{2.10}$$

- H usually denotes a square matrix with the form:

$$H = \begin{pmatrix} 0 & 1 & 0 & \cdots & 0 \\ 0 & 0 & 1 & \cdots & 0 \\ \vdots & \vdots & \vdots & \ddots & \vdots \\ 0 & 0 & 0 & \cdots & 1 \\ 0 & 0 & 0 & \cdots & 0 \end{pmatrix} \tag{2.11}$$

- $\vec{H}_q$ usually denotes a square matrix with the form:

$$\vec{H}_q = \begin{pmatrix} O & I_{m-q} \\ O & O \end{pmatrix} \tag{2.12}$$

where the zero parts are denoted as O with proper dimensions. Obviously, $\vec{H}_q = H^q$.

- $\overleftarrow{H}_q$ usually denotes a square matrix with the form:

$$\overleftarrow{H}_q = \begin{pmatrix} \overleftarrow{I}_q & O \\ O & O \end{pmatrix} \tag{2.13}$$

where $\overleftarrow{I}_q$ is a q-dimensional square matrix:

$$\overleftarrow{I}_q = \begin{pmatrix} 0 & \cdots & 0 & 1 \\ 0 & \cdots & 1 & 0 \\ \vdots & \ddots & \vdots & \vdots \\ 1 & \cdots & 0 & 0 \end{pmatrix} \tag{2.14}$$

2.2 Elementary operation rules

Operation rules of block pulse series are involved in many papers, e.g. Rao and Srinivasan (1978), Kung and Chen (1978), Shih (1978), Hsu and Cheng (1981), Sun (1981), Palanisamy and Bhattacharya (1981), Chen and Jeng (1981), Zeng (1981), Chen and Meng (1982), Wang and Shih (1982), Palanisamy (1983). In this section, we discuss some elementary operation rules of block pulse series. For clarity of expressions, these operation rules are described both in concrete summation forms and in compact vector forms.

1. Constant function. For a real constant function $f(t) = k$, its block pulse series can be simply expanded as:

$$k = k \sum_{i=1}^{m} \phi_i(t) \tag{2.15}$$

Expressing in vector form, it is:

$$\begin{aligned} k &= \begin{pmatrix} k & k & \cdots & k \end{pmatrix} \Phi(t) \\ &= k E^T \Phi(t) \end{aligned} \tag{2.16}$$

Equation (2.15) also implies that the sum of all block pulse functions equals one.

2. Addition and subtraction of functions. For functions $f(t)$ and $g(t)$, their sum or difference can be expanded directly into block pulse series if we rearrange the terms in the results:

$$\begin{aligned} f(t) \pm g(t) &\doteq \sum_{i=1}^{m} f_i \phi_i(t) \pm \sum_{j=1}^{m} g_j \phi_j(t) \\ &= \sum_{i=1}^{m} (f_i \pm g_i) \phi_i(t) \end{aligned} \tag{2.17}$$

Expressing in vector form, it is:

$$\begin{aligned} f(t) \pm g(t) &\doteq \begin{pmatrix} f_1 \pm g_1 & f_2 \pm g_2 & \cdots & f_m \pm g_m \end{pmatrix} \Phi(t) \\ &= \left(F^T \pm G^T \right) \Phi(t) \end{aligned} \tag{2.18}$$

Equation (2.18) shows that the block pulse coefficients of the sum or difference of functions are the sums or differences of the block pulse coefficients of these functions in the same subintervals.

3. Function multiplied by a scalar. If a function $f(t)$ is multiplied by a real scalar k, we have the block pulse series:

$$\begin{aligned} k f(t) &\doteq \sum_{i=1}^{m} (k f_i) \phi_i(t) \\ &= k \sum_{i=1}^{m} f_i \phi_i(t) \end{aligned} \tag{2.19}$$

Expressing in vector form, it is:

$$\begin{aligned} k f(t) &\doteq k \begin{pmatrix} f_1 & f_2 & \cdots & f_m \end{pmatrix} \Phi(t) \\ &= k F^T \Phi(t) \end{aligned} \tag{2.20}$$

Equation (2.20) shows that the block pulse coefficients are magnified k times if a function is multiplied by a scalar k.

4. Multiplication and division of functions. For functions $f(t)$ and $g(t)$, we have the block pulse series of their multiplication:

$$
\begin{aligned}
f(t)g(t) &\doteq \left(\sum_{i=1}^{m} f_i \phi_i(t) \right) \left(\sum_{j=1}^{m} g_j \phi_j(t) \right) \\
&= \sum_{i=1}^{m} f_i g_i \phi_i(t)
\end{aligned}
\tag{2.21}
$$

If $g(t) \neq 0$, we also have the block pulse series of their quotient:

$$
\begin{aligned}
f(t)/g(t) &\doteq \sum_{i=1}^{m} f_i \phi_i(t) \Big/ \sum_{j=1}^{m} g_j \phi_j(t) \\
&\doteq \sum_{i=1}^{m} (f_i/g_i) \phi_i(t)
\end{aligned}
\tag{2.22}
$$

Expressing in vector forms, these equations are:

$$
\begin{aligned}
f(t)g(t) &\doteq \left(\begin{array}{cccc} f_1 g_1 & f_2 g_2 & \cdots & f_m g_m \end{array} \right) \Phi(t) \\
&= F^T D_G \Phi(t) \\
&= G^T D_F \Phi(t) \\
&= E^T D_F D_G \Phi(t)
\end{aligned}
\tag{2.23}
$$

and

$$
\begin{aligned}
f(t)/g(t) &\doteq \left(\begin{array}{cccc} f_1/g_1 & f_2/g_2 & \cdots & f_m/g_m \end{array} \right) \Phi(t) \\
&= F^T D_G^{-1} \Phi(t)
\end{aligned}
\tag{2.24}
$$

where D_F and D_G are diagonal matrices related to the block pulse coefficient vectors F and G, respectively. Here, all the joint terms in the multiplication and division of block pulse series disappear owing to the disjointness of block pulse functions. Equations (2.23) and (2.24) show that the block pulse coefficients of the product and quotient of functions are only the products and quotients of block pulse coefficients of these functions in the same subintervals.

By the way, we also mention here shortly the multiplication of two block pulse vectors. There are two different cases of this multiplication. In the first case, we obtain a matrix:

$$
\Phi(t)\Phi^T(t) = \begin{pmatrix} \phi_1(t) & 0 & \cdots & 0 \\ 0 & \phi_2(t) & \cdots & 0 \\ \vdots & \vdots & \ddots & \vdots \\ 0 & 0 & \cdots & \phi_m(t) \end{pmatrix}
\tag{2.25}
$$

In fact, this expression has the same meaning as the disjointness of block pulse functions which was discussed in (1.4). The matrix in (2.25) is usually called the multiplication matrix of block pulse functions in the literature. It is a diagonal matrix with m block pulse functions as its diagonal entries. Obviously, we have:

$$
F^T \Phi(t)\Phi^T(t) = \Phi^T(t) D_F
\tag{2.26}
$$

In the second case, we obtain a scalar:

$$\Phi^T(t)\Phi(t) = 1 \tag{2.27}$$

In fact, this is just the vector form of the sum of block pulse functions which was discussed in (2.15), because

$$\sum_{i=1}^{m}\left(\phi_i(t)\right)^2 = \sum_{i=1}^{m}\phi_i(t) \tag{2.28}$$

5. Integration and derivation of functions. In order to expand the integral of a function into its block pulse series, we first consider the integral of each single block pulse function $\phi_i(t)$. This integration can be divided into three cases. For the case of $t \in [0,(i-1)h)$, we have:

$$\int_0^t \phi_i(t)dt = 0 \tag{2.29}$$

For the case of $t \in [(i-1)h, ih)$, we have:

$$\begin{aligned}\int_0^t \phi_i(t)dt &= \int_0^{(i-1)h}\phi_i(t)dt + \int_{(i-1)h}^t \phi_i(t)dt \\ &= t - (i-1)h \end{aligned} \tag{2.30}$$

And for the case of $t \in [ih, T)$, we have:

$$\begin{aligned}\int_0^t \phi_i(t)dt &= \int_0^{(i-1)h}\phi_i(t)dt + \int_{(i-1)h}^{ih}\phi_i(t)dt + \int_{ih}^t \phi_i(t)dt \\ &= h \end{aligned} \tag{2.31}$$

As an example, the integrals of a block pulse function set with $m = 4$ are illustrated in Figure 2.2.

The result of these three cases can also be expanded into block pulse series:

$$\int_0^t \phi_i(t)dt \doteq \frac{h}{2}\phi_i(t) + h\sum_{j=i+1}^{m}\phi_j(t) \tag{2.32}$$

and it has the vector form:

$$\int_0^t \phi_i(t)dt \doteq \left(\begin{array}{ccccccc} 0 & \cdots & 0 & h/2 & h & \cdots & h \end{array}\right)\Phi(t) \tag{2.33}$$
$$\uparrow$$
$$i\text{th-entry}$$

In (2.33), the block pulse coefficient vector is rather regular, i.e. the ith entry has the value $h/2$, the entries before it all have the value zero and the entries after it all have the value h.

From the above discussion, the block pulse series of the integrals of all the m block pulse functions can be written together in a compact form:

$$\int_0^t \Phi(t)dt \doteq P\Phi(t) \tag{2.34}$$

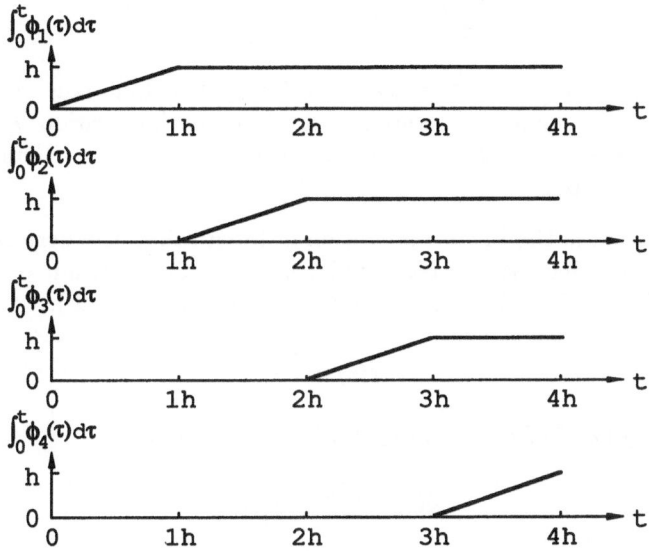

Figure 2.2: Integral of block pulse functions, $(m = 4)$.

where the matrix:

$$P = \frac{h}{2}\begin{pmatrix} 1 & 2 & 2 & \cdots & 2 \\ 0 & 1 & 2 & \cdots & 2 \\ 0 & 0 & 1 & \cdots & 2 \\ \vdots & \vdots & \vdots & \ddots & \vdots \\ 0 & 0 & 0 & \cdots & 1 \end{pmatrix} \qquad (2.35)$$

is defined as the block pulse operational matrix for integration, or simply as the integration operational matrix. This matrix has a regular form, i.e. it is an upper triangular matrix, and its kth row can be obtained by shifting the first row $(k-1)$ positions to the right. We can also verify that all the m eigenvalues of this upper triangular matrix are $h/2$.

Based on this integration operational matrix, it is easy to express the integral of a function $f(t)$ into its block pulse series:

$$\begin{aligned} \int_0^t f(t)dt &\doteq \int_0^t F^T \Phi(t)dt \\ &\doteq F^T P \Phi(t) \end{aligned} \qquad (2.36)$$

If we express the matrix multiplication in (2.36) concretely, we can obtain another form of the block pulse series of integration of functions which also appears frequently in the literature:

$$\int_0^t f(t)dt \doteq \sum_{i=1}^m \left(\frac{h}{2}f_i + h\sum_{j=1}^{i-1} f_j \right) \phi_i(t)$$

$$= \sum_{i=1}^{m} f_i \left(\frac{h}{2} \phi_i(t) + h \sum_{j=i+1}^{m} \phi_j(t) \right) \tag{2.37}$$

Equations (2.36) and (2.37) show that the integral of functions can be transformed approximately into algebraic operations based on the integration operational matrix, so that certain problems involving integrations can be solved in a simpler way through block pulse series expansions.

Since derivation is the inverse operation of integration, the derivative of a function can easily be expanded into its block pulse series from the result of integral obtained above. If $g(t)$ is a differentiable function in the interval $t \in [0, T)$, we denote:

$$f(t) = \frac{dg(t)}{dt} \tag{2.38}$$

After integrating (2.38) from 0 to t on both sides, we have:

$$g(t) - g_0^{(0)} = \int_0^t f(t) dt \tag{2.39}$$

where the initial value of $g(t)$ is denoted by:

$$g(0) = g_0^{(0)} \tag{2.40}$$

Substituting (2.16) and (2.36) into (2.39), we have:

$$\left(G^T - g_0^{(0)} E^T \right) \Phi(t) \doteq F^T P \Phi(t) \tag{2.41}$$

Equating the coefficients of $\phi_i(t)$ ($i = 1, 2, \ldots, m$) on both sides of the above equation, we obtain:

$$G^T - g_0^{(0)} E^T \doteq F^T P \tag{2.42}$$

Noticing that the matrix P is not singular and its inverse P^{-1} exists, the above equation can be rewritten as:

$$F^T \doteq G^T P^{-1} - g_0^{(0)} E^T P^{-1} \tag{2.43}$$

Therefore the block pulse series of the derivative of function $g(t)$ is:

$$\frac{dg(t)}{dt} \doteq \left(G^T P^{-1} - g_0^{(0)} E^T P^{-1} \right) \Phi(t) \tag{2.44}$$

Equation (2.44) shows that the derivative of functions can be transformed approximately into algebraic operations based on the inverse of the integration operational matrix.

Like the integration operational matrix P, the inverse matrix P^{-1} has also a regular upper triangular form:

$$P^{-1} = \frac{2}{h} \begin{pmatrix} 1 & -2 & 2 & \cdots & (-1)^{m-1}2 \\ 0 & 1 & -2 & \cdots & (-1)^{m-2}2 \\ 0 & 0 & 1 & \cdots & (-1)^{m-3}2 \\ \vdots & \vdots & \vdots & \ddots & \vdots \\ 0 & 0 & 0 & \cdots & 1 \end{pmatrix} \tag{2.45}$$

and all its m eigenvalues are $2/h$.

Now we consider the block pulse series of the multiple integral of a function. It is rather simple to derive the related formula because we can apply the integral relation (2.36) successively:

$$\underbrace{\int_0^t \cdots \int_0^t}_{k \text{ times}} f(t)\, dt \cdots dt$$

$$= \int_0^t \cdots \left(\int_0^t \left(\int_0^t f(t) dt \right) dt \right) \cdots dt$$

$$\doteq \int_0^t \cdots \left(\int_0^t F^T P \Phi(t) dt \right) \cdots dt$$

$$\vdots$$

$$\doteq \int_0^t F^T P^{k-1} \Phi(t) dt$$

$$\doteq F^T P^k \Phi(t) \tag{2.46}$$

To obtain the block pulse series of the k times derivative of a function, a similar derivation as in the one time derivative case can be done, only the relation (2.36) should be applied here successively. If $g(t)$ is k times differentiable in the interval $t \in [0, T)$, we denote:

$$f(t) = \frac{d^k g(t)}{dt^k} \tag{2.47}$$

After integrating this equation k times from 0 to t on both sides, we have:

$$\underbrace{\int_0^t \cdots \int_0^t}_{k \text{ times}} f(t)\, dt \cdots dt$$

$$= g(t) - g_0^{(0)} - g_0^{(1)} \int_0^t dt - \cdots - g_0^{(k-1)} \underbrace{\int_0^t \cdots \int_0^t}_{(k-1) \text{ times}} dt \cdots dt \tag{2.48}$$

where $g_0^{(0)}$ is the initial value of $g(t)$, and $g_0^{(i)}$ $(i = 1, 2, \ldots, k-1)$ are the initial values of the successive derivatives of $g(t)$:

$$g_0^{(i)} = \left. \frac{d^i g(t)}{dt^i} \right|_{t=0} \tag{2.49}$$

Since the relation:

$$g_0^{(i)} \underbrace{\int_0^t \cdots \int_0^t}_{i \text{ times}} dt \cdots dt \doteq g_0^{(i)} E^T P^i \Phi(t) \tag{2.50}$$

holds for $i = 1, 2, \ldots, k-1$ according to (2.46), we can straightforwardly obtain:

$$F^T \doteq G^T P^{-k} - g_0^{(0)} E^T P^{-k} - g_0^{(1)} E^T P^{-(k-1)} - \cdots - g_0^{(k-1)} E^T P^{-1} \tag{2.51}$$

from (2.48) and (2.50), where the power matrices P^i ($i = 1, 2, \ldots, k$) are not singular. Therefore the block pulse series of the k times derivative of $g(t)$ is:

$$\frac{d^k g(t)}{dt^k} \doteq \left(G^T P^{-k} - \sum_{i=0}^{k-1} g_0^{(i)} E^T P^{-(k-i)} \right) \Phi(t) \tag{2.52}$$

Above we discussed the operation rules of block pulse series for integrals and derivatives. Theoretically, both these rules can be applied in the block pulse series expansions. But in practice, we do not prefer to use the matrices P^{-k} ($k = 1, 2, \ldots$) because these matrices lead sometimes to divergent results in numerical evaluations, especially when the power k is large. In fact, if problems contain derivative operations, e.g. differential equations, we usually transform them first into integral equation forms, and then apply the block pulse series expansions to obtain numerical results. The details of the manipulation of differential equations will be discussed in Chapters 7 and 8.

6. Convolution integral of functions. In order to expand the convolution integral of functions into block pulse series, we first consider the block pulse series of the convolution integral of two single block pulse functions. Since the Laplace transform of $\phi_i(t) * \phi_j(t)$ is:

$$\begin{aligned}
\mathcal{L}\left\{\phi_i(t) * \phi_j(t)\right\} &= \mathcal{L}\left\{\phi_i(t)\right\} \mathcal{L}\left\{\phi_j(t)\right\} \\
&= \left(\frac{1}{s} e^{-(i-1)hs} \left(1 - e^{-hs}\right)\right) \left(\frac{1}{s} e^{-(j-1)hs} \left(1 - e^{-hs}\right)\right) \\
&= \frac{1}{s^2} e^{-(i+j-2)hs} \left(1 - e^{-hs}\right)^2
\end{aligned} \tag{2.53}$$

its inverse transform yields:

$$\begin{aligned}
&\phi_i(t) * \phi_j(t) \\
&= (t - (i + j - 2)h)\mu(t - (i + j - 2)h) \\
&\quad -2(t - (i + j - 1)h)\mu(t - (i + j - 1)h) \\
&\quad +(t - (i + j)h)\mu(t - (i + j)h) \\
&= \begin{cases} t - (i + j - 2)h & \text{for } (i + j - 2)h \leq t < (i + j - 1)h \\ -t + (i + j)h & \text{for } (i + j - 1)h \leq t < (i + j)h \\ 0 & \text{otherwise} \end{cases}
\end{aligned} \tag{2.54}$$

where $\mu(t)$ is a unit step function:

$$\mu(t) = \begin{cases} 1 & \text{for } t \geq 0 \\ 0 & \text{for } t < 0 \end{cases} \tag{2.55}$$

This convolution integral is illustrated in Figure 2.3.

Based on the block pulse series of $\phi_i(t) * \phi_j(t)$:

$$\phi_i(t) * \phi_j(t) \doteq \sum_{k=1}^{m} \alpha_k \phi_k(t) \tag{2.56}$$

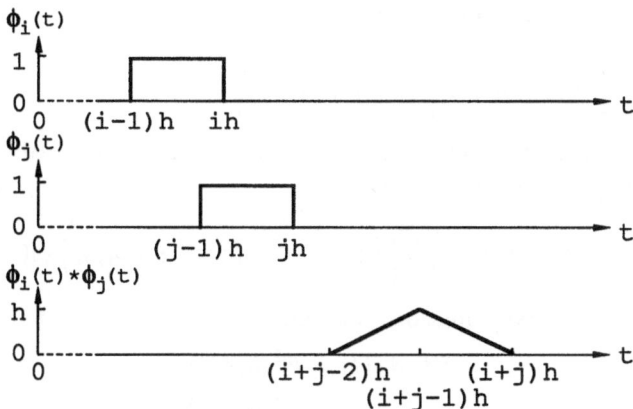

Figure 2.3: Convolution integral of block pulse functions $\phi_i(t) * \phi_j(t)$.

with

$$
\alpha_k = \begin{cases} h/2 & \text{for } k = i + j - 1 \leq m \\ h/2 & \text{for } k = i + j \leq m \\ 0 & \text{otherwise} \end{cases} \tag{2.57}
$$

the block pulse series of the convolution integral $f(t) * g(t)$ can easily be obtained:

$$
\begin{aligned}
c(t) &= f(t) * g(t) \\
&= \int_0^t f(\tau) g(t - \tau) d\tau \\
&\doteq \int_0^t \left(\sum_{i=1}^m f_i \phi_i(\tau) \sum_{j=1}^m g_j \phi_j(t - \tau) \right) d\tau \\
&= \sum_{i=1}^m \sum_{j=1}^m \left(f_i g_j \int_0^t \phi_i(\tau) \phi_j(t - \tau) d\tau \right) \\
&= \sum_{i=1}^m \sum_{j=1}^m f_i g_j \left(\phi_i(t) * \phi_j(t) \right) \\
&\doteq \sum_{i=1}^m \sum_{j=1}^m \sum_{k=1}^m f_i g_j \alpha_k \phi_k(t) \\
&= \frac{h}{2} \sum_{i=1}^m \sum_{j=1}^{m-i+1} f_i g_j \phi_{i+j-1}(t) + \frac{h}{2} \sum_{i=1}^m \sum_{j=1}^{m-i} f_i g_j \phi_{i+j}(t) \tag{2.58}
\end{aligned}
$$

Setting $l = i + j - 1$ for the first term and $l = i + j$ for the second term in the above equation, it can be simplified as:

$$
f(t) * g(t) \doteq \frac{h}{2} \sum_{i=1}^m \sum_{l=i+1}^m f_i (g_{l-i+1} + g_{l-i}) \phi_l(t) + \frac{h}{2} \sum_{i=1}^m f_i g_1 \phi_i(t)
$$

$$= \frac{h}{2} \sum_{l=2}^{m} \sum_{i=1}^{l-1} f_i(g_{l-i+1} + g_{l-i})\phi_l(t) + \frac{h}{2} \sum_{i=1}^{m} f_i g_1 \phi_i(t)$$

$$= \frac{h}{2} \sum_{l=1}^{m} \left(g_1 f_l + \sum_{i=1}^{l-1} (g_{l-i+1} + g_{l-i}) f_i \right) \phi_l(t) \tag{2.59}$$

Since the functions in the convolution integral are interchangeable, the convolution integral $f(t) * g(t)$ can also be expressed as:

$$f(t) * g(t) \doteq \frac{h}{2} \sum_{l=1}^{m} \left(f_1 g_l + \sum_{i=1}^{l-1} (f_{l-i+1} + f_{l-i}) g_i \right) \phi_l(t) \tag{2.60}$$

In vector forms, the above equations become:

$$f(t) * g(t) \doteq F^T J_G \Phi(t)$$
$$\doteq G^T J_F \Phi(t) \tag{2.61}$$

where the matrices J_G and J_F have the forms:

$$J_G = \frac{h}{2} \begin{pmatrix} g_1 & g_1 + g_2 & g_2 + g_3 & \cdots & g_{m-1} + g_m \\ 0 & g_1 & g_1 + g_2 & \cdots & g_{m-2} + g_{m-1} \\ 0 & 0 & g_1 & \cdots & g_{m-3} + g_{m-2} \\ \vdots & \vdots & \vdots & \ddots & \vdots \\ 0 & 0 & 0 & \cdots & g_1 \end{pmatrix} \tag{2.62}$$

and

$$J_F = \frac{h}{2} \begin{pmatrix} f_1 & f_1 + f_2 & f_2 + f_3 & \cdots & f_{m-1} + f_m \\ 0 & f_1 & f_1 + f_2 & \cdots & f_{m-2} + f_{m-1} \\ 0 & 0 & f_1 & \cdots & f_{m-3} + f_{m-2} \\ \vdots & \vdots & \vdots & \ddots & \vdots \\ 0 & 0 & 0 & \cdots & f_1 \end{pmatrix} \tag{2.63}$$

Usually, both matrices J_G and J_F are called the block pulse operational matrices for convolution integral, or simply the convolution operational matrices. They are also upper triangular matrices, and their kth rows can be obtained by shifting the corresponding first rows $(k-1)$ positions to the right, respectively.

By the way, we also mention here shortly the convolution integral of two block pulse vectors. According to (2.57), we obtain a matrix:

$$\Phi(t) * \Phi^T(t) = \begin{pmatrix} \phi_1(t) * \phi_1(t) & \phi_1(t) * \phi_2(t) & \cdots & \phi_1(t) * \phi_m(t) \\ \phi_2(t) * \phi_1(t) & \phi_2(t) * \phi_2(t) & \cdots & \phi_2(t) * \phi_m(t) \\ \vdots & \vdots & \ddots & \vdots \\ \phi_m(t) * \phi_1(t) & \phi_m(t) * \phi_2(t) & \cdots & \phi_m(t) * \phi_m(t) \end{pmatrix}$$

$$\doteq \frac{h}{2} \begin{pmatrix} \phi_1(t) + \phi_2(t) & \phi_2(t) + \phi_3(t) & \cdots & \phi_{m-1}(t) + \phi_m(t) & \phi_m(t) \\ \phi_2(t) + \phi_3(t) & \phi_3(t) + \phi_4(t) & \cdots & \phi_m(t) & 0 \\ \phi_3(t) + \phi_4(t) & \phi_4(t) + \phi_5(t) & \cdots & 0 & 0 \\ \vdots & \vdots & \ddots & \vdots & \vdots \\ \phi_m(t) & 0 & \cdots & 0 & 0 \end{pmatrix} \tag{2.64}$$

which is usually called the convolution matrix of block pulse functions in the literature. We have the relation:

$$\left(F^T\Phi(t)\right) * \left(G^T\Phi(t)\right) = F^T\left(\Phi(t) * \Phi^T(t)\right)G$$
$$\doteq F^T J_G \Phi(t)$$
$$= G^T J_F \Phi(t) \tag{2.65}$$

7. Functions containing time delay $f(t-\tau)$**.** In order to expand a function containing time delay into its block pulse series, we first consider a block pulse function $\phi_i(t)$ containing time delay in two special cases.

In the first case, the time delay is $\tau = qh$ with a nonnegative integer q. Since each block pulse is now shifted merely the distance of q times the width of block pulses in the positive direction, the block pulse function $\phi_i(t-qh)$ can be expressed as:

$$\phi_i(t-qh) = \begin{cases} \phi_{i+q}(t) & \text{for } i \le m-q \\ 0 & \text{for } i > m-q \end{cases} \tag{2.66}$$

Using the notations of (2.7) and (2.11), the above equation can also be written in a vector form:

$$\phi_i(t-qh) = \Delta_i^T H^q \Phi(t) \tag{2.67}$$

In the second case, the time delay is $\tau = \lambda h$ with $0 \le \lambda < 1$, i.e. the time delay is less than the width of block pulses. If we define an auxiliary function of a single block pulse with the width λh:

$$\phi_\lambda(t) = \begin{cases} 1 & \text{for } 0 \le t < \lambda h \\ 0 & \text{otherwise} \end{cases} \tag{2.68}$$

we can compose $\phi_i(t-\lambda h)$ in the following way:

$$\phi_i(t-\tau) = \begin{cases} \phi_i(t) + \phi_\lambda(t-ih) \\ \quad -\phi_\lambda(t-(i-1)h) & \text{for } i=1,2,\ldots,m-1 \\ \phi_i(t) - \phi_\lambda(t-(i-1)h) & \text{for } i=m \end{cases} \tag{2.69}$$

This composition is illustrated in Figure 2.4. After we define an $m \times 1$ vector $\Phi_\lambda(t)$ based on the auxiliary function $\phi_\lambda(t)$:

$$\Phi_\lambda(t) = \left(\begin{array}{cccc} \phi_\lambda(t) & \phi_\lambda(t-h) & \cdots & \phi_\lambda(t-(m-1)h) \end{array} \right)^T \tag{2.70}$$

a block pulse function containing time delay $\tau = \lambda h$ can also be written in a vector form:

$$\phi_i(t-\lambda h) = \Delta_i^T \Phi(t) - \Delta_i^T \Phi_\lambda(t) + \Delta_i^T H \Phi_\lambda(t) \tag{2.71}$$

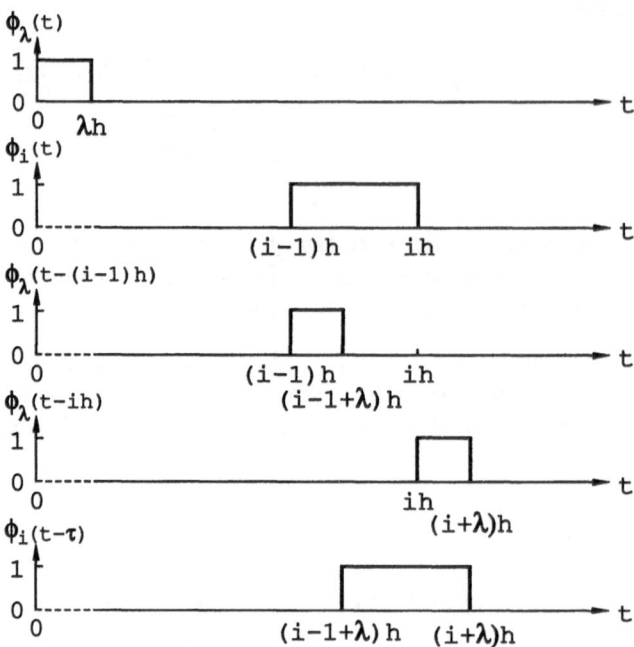

Figure 2.4: Composition of $\phi_i(t - \tau)$, $(\tau = \lambda h)$.

Generally, a block pulse function containing time delay $\tau = (q+\lambda)h$ can be expressed as the combination of the two cases discussed above:

$$\phi_i(t - \tau) = \begin{cases} \phi_{i+q}(t) + \phi_\lambda(t - (i + q)h) \\ \qquad -\phi_\lambda(t - (i + q - 1)h) & \text{for } i < m - q \\ \phi_{i+q}(t) - \phi_\lambda(t - (i + q - 1)h) & \text{for } i = m - q \\ 0 & \text{for } i > m - q \end{cases} \tag{2.72}$$

or in a vector form:

$$\phi_i(t - \tau) = \Delta_i^T H^q \Phi(t) - \Delta_i^T H^q \Phi_\lambda(t) + \Delta_i^T H^{q+1} \Phi_\lambda(t) \tag{2.73}$$

To avoid the expression $\Phi_\lambda(t)$ in the above equation, we expand the function $\phi_i(t - \tau)$ into its block pulse series:

$$\phi_i(t - \tau) = \begin{pmatrix} c_{i,1} & c_{i,2} & \cdots & c_{i,m} \end{pmatrix} \Phi(t) \tag{2.74}$$

where the block pulse coefficients $c_{i,j}$ $(j = 1, 2, \ldots, m)$ are:

$$\begin{aligned} c_{i,j} &= \frac{1}{h} \int_0^T \phi_i(t - \tau)\phi_j(t)dt \\ &= \frac{1}{h} \int_{(j-1)h}^{jh} \phi_i(t - \tau)dt \end{aligned}$$

$$\begin{aligned} &= \frac{1}{h}\Delta_i^T H^q \left(\int_{(j-1)h}^{jh} \Phi(t)dt - \int_{(j-1)h}^{jh} \Phi_\lambda(t)dt + H \int_{(j-1)h}^{jh} \Phi_\lambda(t)dt \right) \\ &= \Delta_i^T \left((1-\lambda)H^q + \lambda H^{q+1} \right) \Delta_j \end{aligned} \tag{2.75}$$

Noticing that the expression $\Delta_i^T \left((1-\lambda)H^q + \lambda H^{q+1} \right) \Delta_j$ is just the single entry positioned in the ith row and jth column of the matrix $(1-\lambda)H^q + \lambda H^{q+1}$, we can expand the whole block pulse function vector containing time delay $\tau = (q+\lambda)h$ into its block pulse series in a vector form:

$$\Phi(t-\tau) = \left((1-\lambda)H^q + \lambda H^{q+1} \right) \Phi(t) \tag{2.76}$$

In the above equation, the matrix $(1-\lambda)H^q + \lambda H^{q+1}$ is usually called the block pulse operational matrix for time delay, or simply the delay operational matrix. Expressing concretely, it is:

$$(1-\lambda)H^q + \lambda H^{q+1} = \begin{array}{c} (q+1)\text{th-column} \\ \downarrow \\ \begin{pmatrix} 0 & \cdots & 0 & 1-\lambda & \lambda & 0 & \cdots & 0 \\ 0 & \cdots & 0 & 0 & 1-\lambda & \lambda & \cdots & 0 \\ \vdots & \cdots & \vdots & \vdots & \vdots & \vdots & \ddots & \vdots \\ 0 & \cdots & 0 & 0 & 0 & 0 & \cdots & \lambda \\ 0 & \cdots & 0 & 0 & 0 & 0 & \cdots & 1-\lambda \\ 0 & \cdots & 0 & 0 & 0 & 0 & \cdots & 0 \\ \vdots & \cdots & \vdots & \vdots & \vdots & \vdots & \cdots & \vdots \\ 0 & \cdots & 0 & 0 & 0 & 0 & \cdots & 0 \end{pmatrix} \end{array} \tag{2.77}$$

According to (2.76), the block pulse series of a function containing time delay $\tau = (q+\lambda)h$ can easily be obtained as:

$$\begin{aligned} f(t-\tau) &\doteq F^T \Phi(t-\tau) \\ &= F^T \left((1-\lambda)H^q + \lambda H^{q+1} \right) \Phi(t) \end{aligned} \tag{2.78}$$

2.3 Other operation rules

In the previous section, some elementary operation rules of block pulse series are discussed. Other operation rules of block pulse series can also be derived similarly. Some of them can even be obtained directly from the elementary operation rules discussed above. Here are some examples.

1. Power of functions. As a special case for multiplication of functions, the block pulse series of the power of a function $f(t)$ can be directly obtained from (2.23):

$$\begin{aligned} f^n(t) &\doteq \left(\begin{array}{cccc} f_1^n & f_2^n & \cdots & f_m^n \end{array} \right) \Phi(t) \\ &= E^T D_F^n \Phi(t) \end{aligned} \tag{2.79}$$

2. Reciprocal of functions. As a special case for division of functions, the block pulse series of the reciprocal of a function $f(t) \neq 0$ can be directly obtained from (2.24):

$$
\begin{aligned}
1/f(t) &\doteq \left(\ 1/f_1 \quad 1/f_2 \quad \cdots \quad 1/f_m\ \right) \Phi(t) \\
&= E^T D_F^{-1} \Phi(t)
\end{aligned}
\tag{2.80}
$$

3. Function of another function. For a function $g(f(t))$, we can first expand it into Taylor series about $f(t) = 0$:

$$
g(f(t)) = g(0) + g^{(1)}(0)f(t) + \frac{1}{2!}g^{(2)}(0)f^2(t) + \cdots + \frac{1}{n!}g^{(n)}(0)f^n(t) + \cdots
\tag{2.81}
$$

and then use the operation rules (2.20) and (2.79) to expand each term of the Taylor series into its corresponding block pulse series, for example:

$$
\frac{1}{n!}g^{(n)}(0)f^n(t) \doteq \sum_{i=1}^{m} \frac{1}{n!}g^{(n)}(0)f_i^n \phi_i
\tag{2.82}
$$

Therefore we have:

$$
g(f(t)) \doteq \sum_{i=1}^{m} \left[g(0) + g^{(1)}(0)f_i + \frac{1}{2!}g^{(2)}(0)f_i^2 + \cdots + \frac{1}{n!}g^{(n)}(0)f_i^n + \cdots \right] \phi_i
\tag{2.83}
$$

Noticing that the term in the square brackets is just the value $g(f_i)$, we can obtain the operation rule:

$$
g(f(t)) \doteq \left(\ g(f_1) \quad g(f_2) \quad \cdots \quad g(f_m)\ \right) \Phi(t)
\tag{2.84}
$$

More generally, this operation rule can be written as:

$$
\begin{aligned}
g\left(f_1(t), \ldots, f_k(t)\right) \doteq \Big(\ & g(f_{1,1}, f_{2,1}, \ldots, f_{k,1}) \\
& g(f_{1,2}, f_{2,2}, \ldots, f_{k,2}) \quad \cdots \quad g(f_{1,m}, f_{2,m}, \ldots, f_{k,m})\ \Big) \Phi(t)
\end{aligned}
\tag{2.85}
$$

4. Integration of function products. For functions $f(t)$ and $g(t)$, the block pulse series of the integral of their product can be directly obtained from (2.23) and (2.36) if the product $f(t)g(t)$ is first manipulated as a whole function $x(t)$:

$$
\begin{aligned}
\int_0^t f(t)g(t)dt &= \int_0^t x(t)dt \\
&\doteq E^T D_X P \Phi(t) \\
&= E^T D_F D_G P \Phi(t)
\end{aligned}
\tag{2.86}
$$

Extending the result of (2.86) to the integral of the product of several functions, we have:

$$
\int_0^t \prod_{i=1}^{k} f_i(t)dt \doteq E^T \prod_{i=1}^{k} D_{F_i} P \Phi(t)
\tag{2.87}
$$

Obviously, the problems about the integral of product of functions can be solved more easily by the block pulse series expansions than by the direct integration.

5. Derivation of function products. Similar to the integration of function products, for functions $f(t)$ and $g(t)$ which are differentiable, the block pulse series of the derivative of their product can be directly obtained from (2.23) and (2.44):

$$\frac{d\,[f(t)g(t)]}{dt} \doteq \left(E^T D_F D_G P^{-1} - f_0^{(0)} g_0^{(0)} E^T P^{-1}\right) \Phi(t) \tag{2.88}$$

where $f_0^{(0)}$ and $g_0^{(0)}$ are the initial values of $f(t)$ and $g(t)$, respectively.

Extending the result of (2.88) to the derivative of the product of several functions, we have:

$$\frac{d\left[\prod_{i=1}^{k} f_i(t)\right]}{dt} \doteq \left(E^T \prod_{i=1}^{k} D_{F_i} P^{-1} - \prod_{i=1}^{k} f_{i,0}^{(0)} E^T P^{-1}\right) \Phi(t) \tag{2.89}$$

6. Backward integration of functions. The backward integration of functions can be manipulated as a special case of the integration. Since the backward integral of the block pulse function vector is:

$$\int_T^t \Phi(t)dt = \int_T^0 \Phi(t)dt + \int_0^t \Phi(t)dt$$
$$\doteq -hE + P\Phi(t)$$
$$= -\frac{h}{2}\begin{pmatrix} 1 & 0 & 0 & \cdots & 0 \\ 2 & 1 & 0 & \cdots & 0 \\ 2 & 2 & 1 & \cdots & 0 \\ \vdots & \vdots & \vdots & \ddots & \vdots \\ 2 & 2 & 2 & \cdots & 1 \end{pmatrix} \Phi(t)$$
$$\doteq -P^T \Phi(t) \tag{2.90}$$

the block pulse series of the backward integral of a function can be expressed as:

$$\int_T^t f(t)dt \doteq \int_T^t F^T \Phi(\tau)d\tau$$
$$\doteq -F^T P^T \Phi(t) \tag{2.91}$$

If we express the matrix multiplication in (2.91) concretely, we can obtain another form of the block pulse series of backward integration of functions which also appears frequently in the literature:

$$\int_T^t f(t)dt \doteq \sum_{i=1}^{m} \left(-\frac{h}{2}f_i - h\sum_{j=i+1}^{m} f_j\right) \phi_i(t)$$
$$= \sum_{i=1}^{m} f_i \left(-\frac{h}{2}\phi_i(t) - h\sum_{j=1}^{i-1} \phi_j(t)\right) \tag{2.92}$$

7. Function containing time shift $f(t+\tau)$. Like the discussions about the functions containing time delay, we first consider a block pulse function $\phi_i(t)$ containing time shift in two special cases.

In the first case, the time shift is $\tau = qh$ with a nonnegative integer q. The block pulse function $\phi_i(t + qh)$ can be expressed as:

$$\phi_i(t + qh) = \begin{cases} \phi_{i-q}(t) & \text{for } i \geq q+1 \\ 0 & \text{for } i < q+1 \end{cases} \tag{2.93}$$

Equation (2.93) can also be written in a vector form:

$$\phi_i(t + qh) = \Delta_i^T \left(H^T\right)^q \Phi(t) \tag{2.94}$$

In the second case, the time shift is $\tau = \lambda h$ with $0 \leq \lambda < 1$, i.e. the time shift is less than the width of block pulses. If we define an auxiliary function of a single block pulse with the width λh:

$$\phi_\lambda^*(t) = \begin{cases} 1 & \text{for } T - \lambda h \leq t < T \\ 0 & \text{otherwise} \end{cases} \tag{2.95}$$

we can compose $\phi_i(t + \lambda h)$ in the following way:

$$\phi_i(t + \tau) = \begin{cases} \phi_i(t) - \phi_\lambda^*(t + (m - i)h) & \text{for } i = 1 \\ \phi_i(t) + \phi_\lambda^*((t - i + 1)h) & \\ \quad - \phi_\lambda^*(t + (m - i)h) & \text{for } i = 2, 3, \ldots, m \end{cases} \tag{2.96}$$

This composition is illustrated in Figure 2.5. After we define an $m \times 1$ vector $\Phi_\lambda^*(t)$ based on the auxiliary function $\phi_\lambda^*(t)$:

$$\Phi_\lambda^*(t) = \left(\phi_\lambda^*(t + (m - 1)h) \quad \phi_\lambda^*(t + (m - 2)h) \quad \cdots \quad \phi_\lambda^*(t) \right)^T \tag{2.97}$$

a block pulse function containing time shift $\tau = \lambda h$ can also be written in a vector form:

$$\phi_i(t + \lambda h) = \Delta_i^T \Phi(t) - \Delta_i^T \Phi_\lambda^*(t) + \Delta_i^T H^T \Phi_\lambda^*(t) \tag{2.98}$$

Generally, a block pulse function containing time shift $\tau = (q + \lambda)h$ can be expressed as the combination of the two cases discussed above:

$$\phi_i(t + \tau) = \begin{cases} 0 & \text{for } i < q+1 \\ \phi_{i-q}(t) - \phi_\lambda^*(t + (m + q - i)h) & \text{for } i = q+1 \\ \phi_{i-q}(t) + \phi_\lambda^*(t + (q - i + 1)h) & \\ \quad - \phi_\lambda^*(t + (m + q - i)h) & \text{for } i > q+1 \end{cases} \tag{2.99}$$

or in a vector form:

$$\phi_i(t + \tau) = \Delta_i^T \left(H^T\right)^q \Phi(t) - \Delta_i^T \left(H^T\right)^q \Phi_\lambda^*(t) + \Delta_i^T \left(H^T\right)^{q+1} \Phi_\lambda^*(t) \tag{2.100}$$

To avoid the expression $\Phi_\lambda^*(t)$ in the above equation, we expand the function $\phi_i(t + \tau)$ into its block pulse series:

$$\phi_i(t + \tau) = \left(c_{i,1} \quad c_{i,2} \quad \cdots \quad c_{i,m} \right) \Phi(t) \tag{2.101}$$

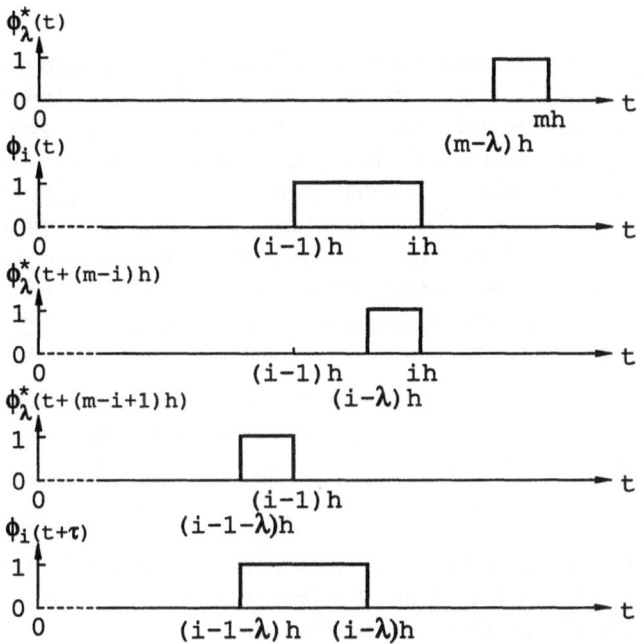

Figure 2.5: Composition of $\phi_i(t + \tau)$, $(\tau = \lambda h)$.

where the block pulse coefficients $c_{i,j}$ $(j = 1, 2, \ldots, m)$ are:

$$
\begin{aligned}
c_{i,j} &= \frac{1}{h} \int_0^T \phi_i(t + \tau)\phi_j(t)dt \\
&= \frac{1}{h} \int_{(j-1)h}^{jh} \phi_i(t + \tau)dt \\
&= \frac{1}{h} \Delta_i^T \left(H^T\right)^q \left(\int_{(j-1)h}^{jh} \Phi(t)dt - \int_{(j-1)h}^{jh} \Phi_\lambda^*(t)dt + H^T \int_{(j-1)h}^{jh} dt \Phi_\lambda^*(t) \right) \\
&= \Delta_i^T \left((1 - \lambda) \left(H^T\right)^q + \lambda \left(H^T\right)^{q+1} \right) \Delta_j
\end{aligned}
\tag{2.102}
$$

Noticing that the expression $\Delta_i^T \left((1 - \lambda) \left(H^T\right)^q + \lambda \left(H^T\right)^{q+1} \right) \Delta_j$ is just the single entry positioned in the ith row and jth column of the matrix $(1 - \lambda) \left(H^T\right)^q + \lambda \left(H^T\right)^{q+1}$, we can expand the whole block pulse function vector containing time shift $\tau = (q + \lambda)h$ into its block pulse series in a vector form:

$$
\Phi(t + \tau) = \left((1 - \lambda) \left(H^T\right)^q + \lambda \left(H^T\right)^{q+1} \right) \Phi(t)
\tag{2.103}
$$

According to (2.103), the block pulse series of a function containing time shift $\tau = (q + \lambda)h$ can easily be obtained as:

$$
f(t + \tau) \doteq F^T \Phi(t + \tau)
$$

$$= F^T \left((1 - \lambda) \left(H^T \right)^q + \lambda \left(H^T \right)^{q+1} \right) \Phi(t) \qquad (2.104)$$

8. Function $f(\tau - t)$. In order to expand a function $f(\tau - t)$ into its block pulse series, we consider first the block pulse series expansion of the function $\phi_i(\tau - t)$ which is related to the single block pulse function $\phi_i(t)$ $(i = 1, 2, \ldots, m)$. But in this discussion, we should modify the definition of block pulse functions a little, i.e. the intervals $t < 0$ and $t \geq T$ are also included in the definition interval now. This modification can as well be expressed by (1.2), which means that each block pulse function has only one rectangular pulse of unit height in a certain small subinterval and takes the value zero in the remaining part of the interval. If the parameter in the variable of the function is $\tau = (q + \lambda)h$ with a nonnegative integer q and a fractional part $0 \leq \lambda < 1$, it is easy to verify that the relation:

$$\phi_i((q + \lambda)h - t) = \begin{cases} \begin{aligned} \phi_{q-i+1}(t) - \phi_\lambda(t - (q - i)h) \\ + \phi_\lambda(t - (q - i + 1)h) \end{aligned} & \text{for } i < q + 1 \\ \phi_\lambda(t - (q - i + 1)h) & \text{for } i = q + 1 \\ 0 & \text{for } i > q + 1 \end{cases} \qquad (2.105)$$

holds in the whole interval $t \in [0, T)$ with only the exception of a finite number of separate points $t = (q - i)h + \lambda h$ and $t = (q - i + 1)h + \lambda h$, where $\phi_\lambda(t)$ is an auxiliary function defined in (2.68). From the relation:

$$\Delta_i^T \overleftarrow{H}_q = \Delta_{q-i+1}^T \qquad (2.106)$$

the function $\phi_i((q + \lambda)h - t)$ can also be written in a vector form:

$$\phi_i((q + \lambda)h - t) = \Delta_i^T \overleftarrow{H}_q \, \Phi(t) - \Delta_i^T \overleftarrow{H}_q \, \Phi_\lambda(t) + \Delta_i^T \overleftarrow{H}_{q+1} \, \Phi_\lambda(t) \qquad (2.107)$$

To avoid the expression $\Phi_\lambda(t)$ in the above equation, we expand the function $\phi_i(\tau - t)$ into its block pulse series:

$$\phi_i(\tau - t) = \left(\begin{array}{cccc} c_{i,1} & c_{i,2} & \cdots & c_{i,m} \end{array} \right) \Phi(t) \qquad (2.108)$$

where the block pulse coefficients $c_{i,j}$ $(j = 1, 2, \ldots, m)$ are:

$$\begin{aligned} c_{i,j} &= \frac{1}{h} \int_0^T \phi_i(\tau - t) \phi_j(t) dt \\ &= \frac{1}{h} \int_{(j-1)h}^{jh} \phi_i(\tau - t) dt \\ &= \frac{1}{h} \Delta_i^T \left(\overleftarrow{H}_q \int_{(j-1)h}^{jh} \Phi(t) dt - \overleftarrow{H}_q \int_{(j-1)h}^{jh} \Phi_\lambda(t) dt + \overleftarrow{H}_{q+1} \int_{(j-1)h}^{jh} \Phi_\lambda(t) dt \right) \\ &= \Delta_i^T \left((1 - \lambda) \overleftarrow{H}_q + \lambda \overleftarrow{H}_{q+1} \right) \Delta_j \end{aligned} \qquad (2.109)$$

Since the expression $\Delta_i^T \left((1 - \lambda) \overleftarrow{H}_q + \lambda \overleftarrow{H}_{q+1} \right) \Delta_j$ is just the single entry positioned in the ith row and jth column of the matrix $(1 - \lambda) \overleftarrow{H}_q + \lambda \overleftarrow{H}_{q+1}$, we can expand the whole function vector:

$$\Phi(\tau - t) = \left(\begin{array}{cccc} \phi_1(\tau - t) & \phi_2(\tau - t) & \cdots & \phi_m(\tau - t) \end{array} \right)^T \qquad (2.110)$$

into its block pulse series in a vector form:

$$\Phi(\tau - t) = \left((1 - \lambda) \overleftarrow{H}_q + \lambda \overleftarrow{H}_{q+1} \right) \Phi(t) \tag{2.111}$$

According to (2.111), the function $f(\tau - t)$ can easily be expanded into its block pulse series:

$$
\begin{aligned}
f(\tau - t) &\doteq F^T \Phi(\tau - t) \\
&= F^T \left((1 - \lambda) \overleftarrow{H}_q + \lambda \overleftarrow{H}_{q+1} \right) \Phi(t)
\end{aligned} \tag{2.112}
$$

Above, we discussed some operation rules of block pulse series. Since these rules are very straightforward, we will not explain their uses through examples. The purpose of the following simple examples is only to show some general phenomena when these operation rules are used.

Example 2.1 Find the block pulse series of double integral of the function $f(t) = t^2$ in the interval $t \in [0, 1)$ with $m = 4$ and $m = 20$ respectively, using the operation rule of integral.

For $m = 4$, the width of block pulses is $h = 0.25$. We first expand the function $f(t)$ into its block pulse series:

$$
\begin{aligned}
f(t) &\doteq F^T \Phi(t) \\
&= \left(\begin{array}{cccc} 0.0208 & 0.1458 & 0.3958 & 0.7708 \end{array} \right) \Phi(t)
\end{aligned}
$$

and then use the integral operation rule (2.46) to obtain the result:

$$
\begin{aligned}
\int_0^t \int_0^t f(t) dt dt &\doteq F^T P^2 \Phi(t) \\
&= \left(\begin{array}{cccc} 0.0003 & 0.0036 & 0.0179 & 0.0589 \end{array} \right) \Phi(t)
\end{aligned} \tag{2.113}
$$

For $m = 20$, the width of block pulses is $m = 0.05$. The same procedure gives the block pulse series:

$$
\begin{aligned}
&\int_0^t \int_0^t f(t) dt dt \\
&\doteq \big(\ 5.2083 \times 10^{-7} \quad 5.7292 \times 10^{-6} \quad 2.8646 \times 10^{-5} \quad 9.4271 \times 10^{-5} \\
&\qquad 2.4010 \times 10^{-4} \quad 5.1615 \times 10^{-4} \quad 9.8490 \times 10^{-4} \quad 1.7214 \times 10^{-3} \\
&\qquad 2.8130 \times 10^{-3} \quad 4.3599 \times 10^{-3} \quad 6.4745 \times 10^{-3} \quad 9.2818 \times 10^{-3} \\
&\qquad 1.2919 \times 10^{-2} \quad 1.7537 \times 10^{-2} \quad 2.3297 \times 10^{-2} \quad 3.0376 \times 10^{-2} \\
&\qquad 3.8959 \times 10^{-2} \quad 4.9247 \times 10^{-2} \quad 6.1454 \times 10^{-2} \quad 7.5803 \times 10^{-2} \ \big) \Phi(t) \quad (2.114)
\end{aligned}
$$

Together with the analytical solution, the block pulse series in (2.113) and (2.114) are illustrated in Figure 2.6. It is clear that the width of block pulses influences the accuracy

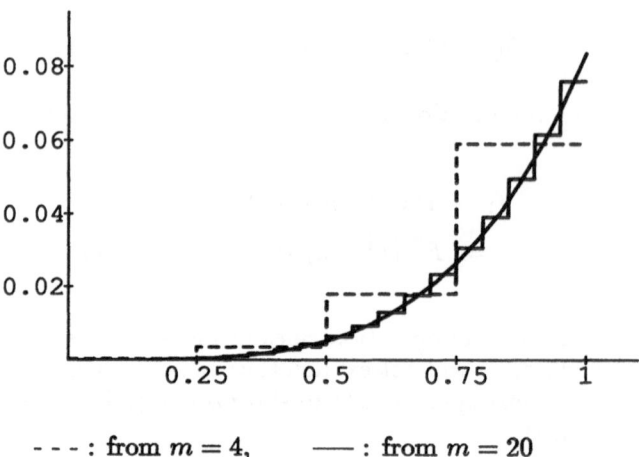

- - - : from $m = 4$,   ——— : from $m = 20$

Figure 2.6: Influence of the width of block pulses.

of the approximations, i.e. a smaller width of block pulses gives a better block pulse series approximation in the results.

If we use the formula (1.14) to evaluate the exact block pulse coefficients of the analytical result $t^4/12$, we obtain for $m = 4$:

$$\int_0^t \int_0^t f(t)dtdt \doteq \left(\begin{array}{cccc} 0.0001 & 0.0020 & 0.0137 & 0.0508 \end{array} \right) \Phi(t) \qquad (2.115)$$

and for $m = 20$:

$$\int_0^t \int_0^t f(t)dtdt$$

$$\doteq \left(\begin{array}{cccc} 1.0417 \times 10^{-7} & 3.2292 \times 10^{-6} & 2.1979 \times 10^{-5} & 8.1354 \times 10^{-5} \end{array} \right.$$

$$2.1885 \times 10^{-4} \quad 4.8448 \times 10^{-4} \quad 9.4073 \times 10^{-4} \quad 1.6626 \times 10^{-3}$$

$$2.7376 \times 10^{-3} \quad 4.2657 \times 10^{-3} \quad 6.3595 \times 10^{-3} \quad 9.1439 \times 10^{-3}$$

$$1.2756 \times 10^{-2} \quad 1.7347 \times 10^{-2} \quad 2.3078 \times 10^{-2} \quad 3.0125 \times 10^{-2}$$

$$\left. 3.8675 \times 10^{-2} \quad 4.8928 \times 10^{-2} \quad 6.1097 \times 10^{-2} \quad 7.5406 \times 10^{-2} \right) \Phi(t) \quad (2.116)$$

We notice that the values in (2.115) and (2.116) are not the same as those in (2.113) and (2.114), respectively. These differences indicate that the block pulse series of the results which are obtained from the operation rules have no more minimal mean square errors between the analytical solution and its piecewise constant approximations, as defined in (1.11).

Example 2.2 Find the block pulse series of the second derivative of the function $f(t) = t^2$, in the interval $t \in [0,1)$ with $m = 4$ and $m = 20$ respectively, using the operation rule of derivatives.

The same procedure of Example 2.1 can be used here. But instead of the matrix P^2 of the integral operation rule (2.46), we apply now the matrix P^{-2} of the derivative operation rule (2.52). Since the initial values in this problem are $f(0) = 0$ and $f^{(1)}(0) = 0$, the derivative operation rule can further be simplified to only one term:

$$f^{(2)}(t) \doteq F^T P^{-2} \Phi(t)$$

For $m = 4$, the block pulse series of the result is:

$$f^{(2)}(t) \doteq \left(\begin{array}{cccc} 1.3333 & 4.0000 & -1.3333 & 6.6667 \end{array} \right) \Phi(t)$$

and $m = 20$, the block pulse series of the result is:

$$f^{(2)}(t) \doteq \left(\begin{array}{ccccc} 1.3333 & 4.0000 & -1.3333 & 6.6667 & -4.0000 \\ 9.3333 & -6.6667 & 12.0000 & -9.3333 & 14.6667 \\ -12.0000 & 17.3333 & -14.6667 & 20.0000 & -17.3333 \\ 22.6667 & -20.0000 & 25.3333 & -22.6667 & 28.0000 \end{array} \right) \Phi(t)$$

Figure 2.7 shows that the derivative operation rule may sometimes lead to divergent results. In contrast with the integral cases in Example 2.1, a smaller width of block pulses gives a stronger divergent tendency in the incorrect result.

As an emphasis, the divergent results in Example 2.2 can remind us to take care if the matrices P^{-k} ($k = 1, 2, \ldots$) are involved in the numerical evaluations, especially when k is large. But it does not mean that the derivative operation rule of block pulse series always leads to incorrect divergent results. For example, if we apply the derivative operation rule to the one time derivative of $f(t) = t^2$, the results of block pulse series are

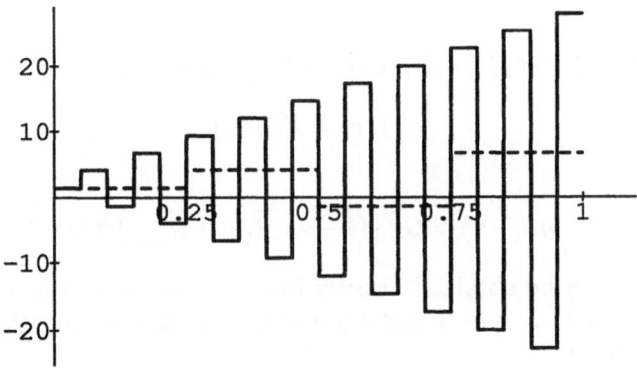

$- - - :$ from $m = 4$, $—— :$ from $m = 20$

Figure 2.7: Divergent results obtained from derivative operation rule.

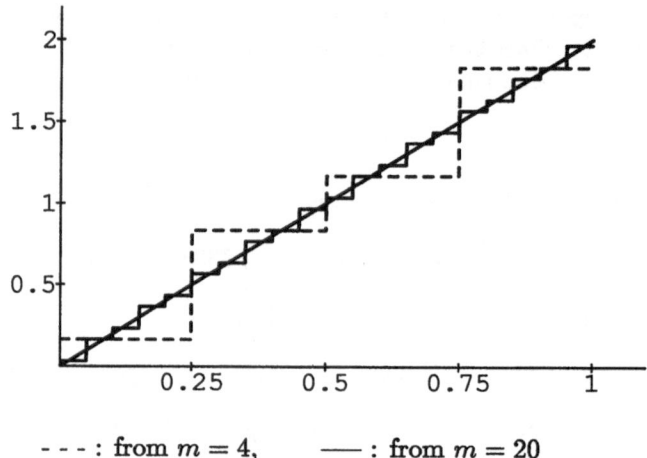

$$- - - : \text{from } m = 4, \qquad \text{——} : \text{from } m = 20$$

Figure 2.8: Convergent results obtained from derivative operation rule.

reasonable. With the analytical result $f^{(1)}(t) = 2t$, the piecewise constant approximate results of both $m = 4$ and $m = 20$ are illustrated in Figure 2.8.

Another thing should also be mentioned. Since all the operation rules of block pulse series are essentially approximate operations, different errors may be introduced in the results of the same problem through different ways of using these operation rules. Here as an example, we derive the operation rule of block pulse series for convolution integral in another way to show such differences.

For the convolution integral:

$$c(t) = \int_0^t f(\tau)g(t - \tau)d\tau \qquad (2.117)$$

we first expand $f(\tau)$ and $g(t - \tau)$ into their block pulse series:

$$f(\tau) \doteq F^T \Phi(\tau) \qquad (2.118)$$

and

$$g(t - \tau) \doteq G^T \left((1 - \lambda) \overleftarrow{H}_q + \lambda \overleftarrow{H}_{q+1} \right) \Phi(\tau) \qquad (2.119)$$

Equation (2.119) can be obtained directly from the operation rule (2.112), where $t = (q + \lambda)h$ with $q = 0, 1, \ldots, m - 1$ and $0 \le \lambda < 1$. Inserting these both block pulse series into (2.117), we have:

$$
\begin{aligned}
c((q + \lambda)h) & \doteq \int_0^{(q+\lambda)h} F^T \Phi(\tau)\Phi^T(\tau) \left((1 - \lambda) \overleftarrow{H}_q + \lambda \overleftarrow{H}_{q+1} \right)^T G d\tau \\
& = \left(\int_0^{(q+\lambda)h} \Phi^T(\tau)d\tau \right) D_F \left((1 - \lambda) \overleftarrow{H}_q + \lambda \overleftarrow{H}_{q+1} \right)^T G \qquad (2.120)
\end{aligned}
$$

Since the integral part is:

$$\int_0^{(q+\lambda)h} \Phi^T(\tau)d\tau = h\left(\begin{array}{ccccccc} 1 & \cdots & 1 & \lambda & 0 & \cdots & 0 \end{array}\right)$$

$$\begin{array}{c} \uparrow \\ (q+1)\text{th-entry} \end{array}$$

(2.121)

the equation (2.120) becomes:

$$c((q+\lambda)h)$$
$$\doteq h\left(\begin{array}{ccccccc} f_1 & \cdots & f_q & \lambda f_{q+1} & 0 & \cdots & 0 \end{array}\right)\left((1-\lambda)\overleftarrow{H}_q + \lambda\overleftarrow{H}_{q+1}\right)^T G \qquad (2.122)$$

From (2.122), the block pulse coefficients of the convolution integral (2.117) can be evaluated according to (1.14):

$$\begin{aligned}
c_{q+1} &= \frac{1}{h}\int_0^1 c((q+\lambda)h)d\lambda h \\
&= h\left(\begin{array}{cccccc} \frac{1}{2}f_1 & \cdots & \frac{1}{2}f_q & \frac{1}{6}f_{q+1} & 0 & \cdots & 0 \end{array}\right)\overleftarrow{H}_q^T G \\
&\quad + h\left(\begin{array}{cccccc} \frac{1}{2}f_1 & \cdots & \frac{1}{2}f_q & \frac{1}{3}f_{q+1} & 0 & \cdots & 0 \end{array}\right)\overleftarrow{H}_{q+1}^T G \\
&= h\left(\begin{array}{ccccccc} \frac{1}{2}f_q & \cdots & \frac{1}{2}f_1 & 0 & 0 & \cdots & 0 \end{array}\right)G \\
&\quad + h\left(\begin{array}{ccccccc} \frac{1}{2}f_{q+1} & \cdots & \frac{1}{2}f_2 & \frac{1}{3}f_1 & 0 & \cdots & 0 \end{array}\right)G \\
&= \frac{h}{3}f_1 g_{q+1} + \frac{h}{2}\sum_{i=1}^q g_i(f_{q-i+1} + f_{q-i+2}) \qquad (2.123)
\end{aligned}$$

where $q = 0, 1, \ldots, m-1$. Therefore the block pulse series of the convolution integral can be expressed in the vector form:

$$\int_0^t f(\tau)g(t-\tau)d\tau \doteq G^T \tilde{J}_F \Phi(t) \qquad (2.124)$$

with

$$\tilde{J}_F = \begin{pmatrix}
\frac{h}{3}f_1 & \frac{h}{2}(f_1+f_2) & \frac{h}{2}(f_2+f_3) & \cdots & \frac{h}{2}(f_{m-1}+f_m) \\
0 & \frac{h}{3}f_1 & \frac{h}{2}(f_1+f_2) & \cdots & \frac{h}{2}(f_{m-2}+f_{m-1}) \\
0 & 0 & \frac{h}{3}f_1 & \cdots & \frac{h}{2}(f_{m-3}+f_{m-2}) \\
\vdots & \vdots & \vdots & \ddots & \vdots \\
0 & 0 & 0 & \cdots & \frac{h}{3}f_1
\end{pmatrix} \qquad (2.125)$$

Comparing the upper triangular matrices in (2.63) and (2.125), we can find that the matrix entries positioned on the main diagonal lines are not the same. This derivation shows that different ways of using the operation rules of block pulse series may lead to different results. The errors of the convolution integral rules (2.61) and (2.124) will be discussed later in Example 6.1.

2.4   Recursive formulas of integrals

In the discussions about the operation rule of multiplication in Section 2.2, we mentioned that all the joint terms in the block pulse series disappear owing to the disjointness of block pulse functions. This means that each block pulse coefficient of a product is related only to the block pulse coefficients of the multiplicators in the same single subinterval. This can be seen clearly in (2.23) because the matrices involved in it are diagonal. Similar diagonal matrices also appear in the operation rule of division. But for some other operation rules, e.g. integration, derivation and convolution integral, the evaluations of block pulse coefficients are not so simple. In these cases, each block pulse coefficient in the results is related not only to the block pulse coefficients of the original functions in the same single subinterval, but also to those in other subintervals. In the case of convolution integral, the block pulse coefficients of $c(t) = f(t) * g(t)$ can only be expressed as:

$$c_i = f_i g_1 + \sum_{j=1}^{i-1} f_j(g_{i-j} + g_{i-j+1}) \tag{2.126}$$

or

$$c_i = f_1 g_i + \sum_{j=1}^{i-1} (f_{i-j} + f_{i-j+1}) g_j \tag{2.127}$$

from the regular upper triangular forms of the convolution operational matrices J_G or J_F. Obviously, the previous $(i-1)$ terms should be included in the above sums for evaluating the ith block pulse coefficient.

From the regular upper triangular forms of the integration operational matrix and its powers P^k ($k = 1, 2, \ldots$), we can express the block pulse coefficients of integrals in the sums similar to (2.126) and (2.127). But unlike the convolution integral case, we can also derive some simpler recursive formulas for evaluating the block pulse coefficients of integrals because the integration operational matrix is more regular than the convolution operational matrix. Here are some examples.

For the cases of single and double integrals, we use the notations (2.39) and (2.48) which are written here once more:

$$\int_0^t f(t)dt = g(t) - g_0^{(0)} \tag{2.128}$$

and

$$\int_0^t \int_0^t f(t)dtdt = g(t) - g_0^{(0)} - g_0^{(1)} \int_0^t dt \tag{2.129}$$

where $g_0^{(0)}$ and $g_0^{(1)}$ are initial values as defined in (2.40) and (2.49), respectively. The intention of using these notations is to extend the results directly to the derivatives after the recursive formulas for the integrals are obtained.

In the case of single integral, the block pulse coefficient vectors in (2.128) have the relation:

$$G^T \doteq F^T P + g_0^{(0)} E^T \qquad (2.130)$$

Noticing that the matrices $(I - H)$ and $P(I - H)$ have special forms:

$$I - H = \begin{pmatrix} 1 & -1 & 0 & 0 & \cdots & 0 \\ 0 & 1 & -1 & 0 & \cdots & 0 \\ 0 & 0 & 1 & -1 & \cdots & 0 \\ 0 & 0 & 0 & 1 & \cdots & 0 \\ \vdots & \vdots & \vdots & \vdots & \ddots & \vdots \\ 0 & 0 & 0 & 0 & \cdots & 1 \end{pmatrix} \qquad (2.131)$$

and

$$P(I - H) = \frac{h}{2} \begin{pmatrix} 1 & 1 & 0 & 0 & \cdots & 0 \\ 0 & 1 & 1 & 0 & \cdots & 0 \\ 0 & 0 & 1 & 1 & \cdots & 0 \\ 0 & 0 & 0 & 1 & \cdots & 0 \\ \vdots & \vdots & \vdots & \vdots & \ddots & \vdots \\ 0 & 0 & 0 & 0 & \cdots & 1 \end{pmatrix} \qquad (2.132)$$

a simple equation can be obtained through postmultiplying a matrix $(I - H)$ in (2.130):

$$G^T(I - H) \doteq F^T P(I - H) + g_0^{(0)} E^T(I - H) \qquad (2.133)$$

Equation (2.133) shows that the block pulse coefficients of the integral $g(t)$ can be recursively evaluated from those of $f(t)$:

$$g_i = \begin{cases} \dfrac{h}{2} f_1 + g_0^{(0)} & \text{for } i = 1 \\[2mm] g_{i-1} + \dfrac{h}{2} (f_i + f_{i-1}) & \text{for } i = 2, 3, \ldots, m \end{cases} \qquad (2.134)$$

In the case of double integral, the block pulse coefficient vectors in (2.129) have the relation:

$$G^T \doteq F^T P^2 + g_0^{(0)} E^T + g_0^{(1)} E^T P \qquad (2.135)$$

where the matrix P^2 has a regular form:

$$P^2 = \frac{h^2}{4} \begin{pmatrix} 1 & 4 & 8 & \cdots & 4(m-1) \\ 0 & 1 & 4 & \cdots & 4(m-2) \\ 0 & 0 & 1 & \cdots & 4(m-3) \\ \vdots & \vdots & \vdots & \ddots & \vdots \\ 0 & 0 & 0 & \cdots & 1 \end{pmatrix} \qquad (2.136)$$

Noticing that the matrices $(I-H)^2$, $P(I-H)^2$ and $P^2(I-H)^2$ also have special forms:

$$(I-H)^2 = \begin{pmatrix} 1 & -2 & 1 & 0 & \cdots & 0 \\ 0 & 1 & -2 & 1 & \cdots & 0 \\ 0 & 0 & 1 & -2 & \cdots & 0 \\ 0 & 0 & 0 & 1 & \cdots & 0 \\ \vdots & \vdots & \vdots & \vdots & \ddots & \vdots \\ 0 & 0 & 0 & 0 & \cdots & 1 \end{pmatrix} \qquad (2.137)$$

$$P(I-H)^2 = \frac{h}{2} \begin{pmatrix} 1 & 0 & -1 & 0 & \cdots & 0 \\ 0 & 1 & 0 & -1 & \cdots & 0 \\ 0 & 0 & 1 & 0 & \cdots & 0 \\ 0 & 0 & 0 & 1 & \cdots & 0 \\ \vdots & \vdots & \vdots & \vdots & \ddots & \vdots \\ 0 & 0 & 0 & 0 & \cdots & 1 \end{pmatrix} \qquad (2.138)$$

and

$$P^2(I-H)^2 = \frac{h^2}{4} \begin{pmatrix} 1 & 2 & 1 & 0 & \cdots & 0 \\ 0 & 1 & 2 & 1 & \cdots & 0 \\ 0 & 0 & 1 & 2 & \cdots & 0 \\ 0 & 0 & 0 & 1 & \cdots & 0 \\ \vdots & \vdots & \vdots & \vdots & \ddots & \vdots \\ 0 & 0 & 0 & 0 & \cdots & 1 \end{pmatrix} \qquad (2.139)$$

a simple equation can be obtained through postmultiplying a matrix $(I-H)^2$ in (2.135):

$$G^T(I-H)^2 \doteq F^T P^2 (I-H)^2 + g_0^{(0)} E^T (I-H)^2 + g_0^{(1)} E^T P (I-H)^2 \qquad (2.140)$$

Equation (2.140) shows that the block pulse coefficients of the double integral $g(t)$ can be recursively evaluated from those of $f(t)$:

$$g_i = \begin{cases} \dfrac{h^2}{4} f_1 + g_0^{(0)} + \dfrac{h}{2} g_0^{(1)} & \text{for } i = 1 \\[2mm] 2g_1 + \dfrac{h^2}{4}(f_2 + 2f_1) - g_0^{(0)} + \dfrac{h}{2} g_0^{(1)} & \text{for } i = 2 \\[2mm] 2g_{i-1} - g_{i-2} + \dfrac{h^2}{4}(f_i + 2f_{i-1} + f_{i-2}) & \text{for } i = 3, 4, \ldots, m \end{cases} \qquad (2.141)$$

The recursive formulas discussed above can save computations, if the block pulse coefficients of integrals should be evaluated. This is one advantage of these recursive formulas, especially when the number of block pulse functions is large. Moreover, although (2.134) and (2.141) indicate that the times of recursion must be restricted in $i \leq m$, we should not take care of the value of i in the practical uses of these formulas. We can assume that the number m is large enough so that the times of recursion always satisfies the restriction $i \leq m$. This is another advantage of these recursive formulas.

The recursive formulas for the derivatives can be obtained directly from the results (2.134) and (2.141) because integration and derivation are inverse operations. For example, for the derivative:

$$f(t) = \frac{dg(t)}{dt} \tag{2.142}$$

we can rewrite (2.134) as:

$$f_i = \begin{cases} \dfrac{2}{h}g_1 - \dfrac{2}{h}g_0^{(0)} & \text{for } i = 1 \\[2ex] -f_{i-1} + \dfrac{2}{h}(g_i - g_{i-1}) & \text{for } i = 2, 3, \ldots, m \end{cases} \tag{2.143}$$

But we will not discuss the recursive formulas of the successive derivatives any further, because they are not very practical in numerical evaluations due to the divergence problem.

2.5   Notes on operational matrices

In Section 2.2, we discussed some operation rules of block pulse series, which contain operational matrices. These operations are integration, convolution integral and time delay. As we introduced the block pulse operational matrices there, we first applied the requested operations to the single block pulse functions to obtain the analytical solutions, and then we expanded the analytical solutions into their block pulse series to obtain the entries of the operational matrices. Figures 2.2, 2.3 and 2.4 show clearly how the analytical solutions are approximated by the piecewise constant functions. Now in this section, we use another way (Hwang and Shih, 1986) to construct these same block pulse operational matrices once more. Based on the approximations of the integral and delay operators in the Laplace domain, this new way gives much simpler derivations and reveals the relations between the operational matrices in the block pulse series expansions and the approximate operators in the Laplace transforms.

1.   Integration operational matrix. As well known, the delay operator in the Laplace domain can be approximated as:

$$e^{-hs} \doteq \frac{2 - hs}{2 + hs} \tag{2.144}$$

This approximation is also equivalent to the Tustin integrator:

$$\begin{aligned} \frac{1}{s} &\doteq \frac{h}{2}\frac{1 + e^{-hs}}{1 - e^{-hs}} \\ &= h\left(\frac{1}{2} + e^{-hs} + e^{-2hs} + e^{-3hs} + \cdots\right) \end{aligned} \tag{2.145}$$

In order to introduce the block pulse operational matrix for integration, we apply this approximate integrator to the integral of the ith block pulse function $\phi_i(t)$ $(i = 1, 2, \ldots, m)$:

$$\mathcal{L}\left\{\int_0^t \phi_i(t)dt\right\} = \frac{1}{s}\mathcal{L}\left\{\phi_i(t)\right\}$$

$$\doteq h \left(\frac{1}{2} + e^{-hs} + e^{-2hs} + e^{-3hs} + \cdots \right) \mathcal{L} \{\phi_i(t)\} \tag{2.146}$$

Since the Laplace transform of the ith block pulse function is:

$$\mathcal{L} \{\phi_i(t)\} = \frac{1}{s} \left(e^{-(i-1)hs} - e^{-ihs} \right) \tag{2.147}$$

the expression (2.146) becomes:

$$
\begin{aligned}
\mathcal{L} \left\{ \int_0^t \phi_i(t) dt \right\} &= \frac{h}{2s} \left(e^{-(i-1)hs} - e^{-ihs} \right) \\
&+ \frac{h}{s} \left(e^{-ihs} - e^{-(i+1)hs} \right) + \frac{h}{s} \left(e^{-(i+1)hs} - e^{-(i+2)hs} \right) + \cdots \\
&= \frac{h}{2} \mathcal{L} \{\phi_i(t)\} + h \mathcal{L} \{\phi_{i+1}(t)\} + h \mathcal{L} \{\phi_{i+2}(t)\} + \cdots
\end{aligned} \tag{2.148}
$$

and its inverse Laplace transform is:

$$\int_0^t \phi_i(t) dt \doteq \frac{h}{2} \phi_i(t) + h \phi_{i+1}(t) + h \phi_{i+2}(t) + \cdots \tag{2.149}$$

Noticing that the block pulse functions $\phi_{m+1}(t)$, $\phi_{m+2}(t)$, ... are not defined, they should be truncated in the above equation. Therefore for $\phi_i(t)$ ($i = 1, 2, \ldots, m$), the m equations in the form of (2.149) can be expressed together in a matrix form which is just the operation rule of integration (2.34) based on the operational matrix P.

2. Convolution operational matrix. Since (2.53) gives the Laplace transform of two block pulse functions $\phi_i(t)$ and $\phi_j(t)$ ($i, j = 1, 2, \ldots, m$):

$$\mathcal{L} \{\phi_i(t) * \phi_j(t)\} = \frac{1}{s^2} \left(e^{-(i+j-2)hs} - 2 e^{-(i+j-1)hs} + e^{-(i+j)hs} \right) \tag{2.150}$$

the substitution of the approximate integrator (2.145) into (2.150) yields:

$$
\begin{aligned}
\mathcal{L} \{\phi_i(t) * \phi_j(t)\} &= \frac{h}{2s} \left(e^{-(i+j-2)hs} - e^{-(i+j)hs} \right) \\
&= \frac{h}{2s} \left(e^{-(i+j-2)hs} - e^{-(i+j-1)hs} \right) + \frac{h}{2s} \left(e^{-(i+j-1)hs} - e^{-(i+j)hs} \right) \\
&= \frac{h}{2} \mathcal{L} \{\phi_{i+j-1}(t)\} + \frac{h}{2} \mathcal{L} \{\phi_{i+j}(t)\}
\end{aligned} \tag{2.151}
$$

Its inverse Laplace transform is:

$$\phi_i(t) * \phi_j(t) = \frac{h}{2} \phi_{i+j-1}(t) + \frac{h}{2} \phi_{i+j}(t) \tag{2.152}$$

Similar to the case of the integration operational matrix above, the terms $\phi_{m+1}(t)$, $\phi_{m+2}(t)$, ... in (2.152) should also be truncated when they appear. For $\phi_i(t) * \phi_j(t)$ ($i, j = 1, 2, \ldots, m$), the m^2 equations in the form of (2.152) can be expressed together in a matrix form which is just the convolution matrix of block pulse functions (2.64). Therefore for functions $f(t)$ and $g(t)$, the relation (2.65) gives directly the operation rule of convolution integral (2.61) based on the operational matrices J_G or J_F.

3. Delay operational matrix. For the ith block pulse function $\phi_i(t - \tau)$ where the time delay $\tau = (q + \lambda)h$ with $0 \leq \lambda < 1$, the shift operation of the Laplace transform gives:

$$\mathcal{L}\{\phi_i(t - \tau)\} = e^{-(q+\lambda)hs}\mathcal{L}\{\phi_i(t)\} \tag{2.153}$$

Noticing that the ith block pulse function with q steps delay is the $(i+q)$th block pulse function:

$$e^{-qhs}\mathcal{L}\{\phi_i(t)\} = \mathcal{L}\{\phi_{i+q}(t)\} \tag{2.154}$$

and the delay operator in the Laplace domain with $0 \leq \lambda < 1$ can be approximated by:

$$e^{-\lambda hs} \doteq (1 - \lambda) + \lambda e^{-hs} \tag{2.155}$$

the expression (2.153) becomes:

$$\mathcal{L}\{\phi_i(t - \tau)\} = (1 - \lambda)\mathcal{L}\{\phi_{i+q}(t)\} + \lambda\mathcal{L}\{\phi_{i+q+1}(t)\} \tag{2.156}$$

Its inverse Laplace transform is:

$$\phi_i(t - \tau) = (1 - \lambda)\phi_{i+q}(t) + \lambda\phi_{i+q+1}(t) \tag{2.157}$$

where the terms $\phi_{m+1}(t)$, $\phi_{m+2}(t), \ldots$ should also be truncated when they appear. Therefore for $\phi_i(t)$ $(i = 1, 2, \ldots, m)$, the m equations in the form of (2.157) can be expressed together in a matrix form which is just the operation rule of time delay (2.76) based on the operational matrix $(1 - \lambda)H^q + \lambda H^{q+1}$.

Since the block pulse functions have special forms of (2.147), they can easily be transformed from one to another by the delay operator in the Laplace domain. By this distinct property, the above derivations of the block pulse operational matrices can be done in a uniform way. Among these derivations, the one about the integral operation is especially noticeable. Equation (2.145) indicates that the Tustin integrator can be expressed by the delay operators, and equation (2.148) indicates that the integration operational matrix of block pulse functions has a close relation with the Tustin integrator. This close relation will be utilized to develop the block pulse transform in Chapter 4.

Chapter 3

Block pulse operators

The idea of transformations is used frequently in mathematics. After the transformations are defined appropriately, mathematical expressions and solutions of many problems can be simplified. For example, the solutions of certain differential equations are complicated in the time domain, but after they are transformed into the corresponding algebraic equations according to the rules of the Laplace transformation, their solutions in the Laplace domain are much easier. The idea of transformations can also be applied in the block pulse function technique to simplify expressions. As discussed in the previous chapters, original functions can be expanded into their block pulse series, and operations of original functions can be converted to the operations of their block pulse series. Since block pulse coefficients take informations of the corresponding block pulse series of the original functions, a transformation can be introduced to express the relations between functions and their block pulse coefficients. As proposed by some authors (Wang, 1983; Wang and Jiang, 1984), this transformation is called block pulse operator. Although the essence of this operator is the same as the block pulse series expansion, it gives a proper mathematical frame for the block pulse function technique and is advantageous to the convergence analysis of block pulse series expansions.

3.1   Definition of block pulse operators

Let M^m be an m-dimensional linear normed space with the norm:

$$\|X\| = \left(\sum_{i=1}^{m} x_i^2 \right)^{\frac{1}{2}} \tag{3.1}$$

where $X = (\ x_1\ \ x_2\ \ \cdots\ \ x_m\ ) \in M^m$. Let $L_2[0,T)$ be a Hilbert space of real bounded functions which are square integrable. Its inner product is defined as:

$$< f(t), g(t) > = \int_0^T f(t)g(t)dt \tag{3.2}$$

where $f(t), g(t) \in L_2[0,T)$. From the inner product, the norm is induced as:

$$\|f(t)\| = (< f(t), f(t) >)^{\frac{1}{2}} \tag{3.3}$$

and both M^m and $L_2[0,T)$ are Banach spaces. Let N^m be a linear space spanned by the block pulse function set $\{\phi_i(t); \ i = 1, 2, \ldots, m\}$, or equivalently, spanned by the orthonormal block pulse function set $\{\varphi_i(t); \ i = 1, 2, \ldots, m\}$. Obviously $N^m \in L_2[0,T)$.

According to the orthogonal projection theorem (Curtain and Pritchard, 1977), for an arbitrary function $f(t) \in L_2[0,T)$, there exists a unique $\hat{f}_m(t) \in N^m$:

$$\hat{f}_m(t) \ = \ \sum_{i=1}^{m} <f(t), \varphi_i(t)> \varphi_i(t)$$

$$= \ \frac{1}{h} \sum_{i=1}^{m} <f(t), \phi_i(t)> \phi_i(t) \tag{3.4}$$

such that

$$\left(f(t) - \hat{f}_m(t)\right) \perp N^m \tag{3.5}$$

and

$$\left\| f(t) - \hat{f}_m(t) \right\| = \min_{\tilde{f}_m(t) \in N^m} \left\| f(t) - \tilde{f}_m(t) \right\| \tag{3.6}$$

Comparing (3.2), (3.4) with (1.14), we notice that the projection of function $f(t)$ on the subspace N^m is just the block pulse series of $f(t)$:

$$\hat{f}_m(t) \ = \ \sum_{i=1}^{m} f_i \phi_i(t)$$

$$= \ F^T \Phi(t) \tag{3.7}$$

where the vector $F^T \in M^m$. Equations (3.5) and (3.6) show that the block pulse series $\hat{f}_m(t)$ obtained from the orthogonal projection is the best approximation of $f(t)$ with respect to the subspace N^m. According to the completeness of block pulse functions discussed in Section 1.4, $\hat{f}_m(t)$ satisfies:

$$\lim_{m \to \infty} \left\| f(t) - \hat{f}_m(t) \right\| = 0 \tag{3.8}$$

Based on the above discussions, we introduce the definition of block pulse operator. For a given positive integer m and an arbitrary function $f(t) \in L_2[0,T)$, the block pulse operator $\mathcal{B}$ determines a unique real m-dimensional vector $F^T \in M^m$:

$$\mathcal{B}\{f(t)\} = F^T \tag{3.9}$$

where the vector F is evaluated from:

$$F = \frac{1}{h} \int_0^T f(t) \Phi(t) dt \tag{3.10}$$

In other words, the block pulse operator produces an image F^T of the function $f(t)$ in the block pulse domain. Since the block pulse operator transforms a function $f(t)$ to an m-dimensional vector F^T whose entries are integral mean values of the function in the corresponding subintervals, we can also say that the block pulse operator discretizes continuous functions.

Example 3.1 Find the image of the function $f(t) = t^2$ under the block pulse operator, in the interval $t \in [0, 1)$ with $m = 4$.

According to the definition of the block pulse operator in (3.9), the block pulse coefficients of the function $f(t)$ should first be determined. This evaluation procedure is just the same as in Example 1.1. After all the block pulse coefficients are obtained, the image given by the block pulse operator can be directly expressed as:

$$\mathcal{B}\{f(t)\} = \left(\ 1/48 \quad 7/48 \quad 19/48 \quad 37/48\ \right)$$

Although this solution takes the same meaning as the one in Example 1.1, the expression via the block pulse operator has a more compact form, because the block pulse function vector $\Phi(t)$ does not appear.

3.2 Properties of block pulse operators

The block pulse operator has the following properties which will be applied in the discussions of operation rules in the next section.

1. The operator is linear. For functions $f(t), g(t) \in L_2[0, T]$, we have:

$$
\begin{aligned}
\mathcal{B}\{\alpha f(t) + \beta g(t)\} \\
&= \left(\frac{1}{h}\int_0^T (\alpha f(t) + \beta g(t))\,\Phi(t)dt\right)^T \\
&= \alpha\left(\frac{1}{h}\int_0^T f(t)\Phi(t)dt\right)^T + \beta\left(\frac{1}{h}\int_0^T g(t)\Phi(t)dt\right)^T \\
&= \alpha\mathcal{B}\{f(t)\} + \beta\mathcal{B}\{g(t)\}
\end{aligned}
\tag{3.11}
$$

Therefore the block pulse operator is a linear operator.

2. The operator is continuous. For a function $f(t)$ and a sequence $\{f_n(t)\}$ which are in the space $L_2[0, T]$, suppose that:

$$\lim_{n\to\infty} \|f_n(t) - f(t)\| = 0 \tag{3.12}$$

we have:

$$
\begin{aligned}
\|\mathcal{B}\{f_n(t)\} - \mathcal{B}\{f(t)\}\| \\
&= \left\|F_n^T - F^T\right\| \\
&= \left(\sum_{i=1}^m (f_{n,i} - f_i)^2\right)^{\frac{1}{2}} \\
&= \frac{1}{h}\left(\sum_{i=1}^m \left(\int_0^T (f_n(t) - f(t))\,\phi_i(t)dt\right)^2\right)^{\frac{1}{2}} \\
&= \frac{1}{h}\left(\sum_{i=1}^m \left(\int_{(i-1)h}^{ih} (f_n(t) - f(t))\,dt\right)^2\right)^{\frac{1}{2}}
\end{aligned}
\tag{3.13}
$$

The Schwarz inequality gives:

$$\|\mathcal{B}\left\{f_n(t)\right\} - \mathcal{B}\left\{f(t)\right\}\|$$

$$\leq \frac{1}{h}\left(\sum_{i=1}^{m} h \int_{(i-1)h}^{ih} (f_n(t) - f(t))^2 \, dt\right)^{\frac{1}{2}}$$

$$= \sqrt{\frac{1}{h}} \left(\sum_{i=1}^{m} \int_{(i-1)h}^{ih} (f_n(t) - f(t))^2 \, dt\right)^{\frac{1}{2}}$$

$$= \sqrt{\frac{1}{h}} \left(\int_{0}^{T} (f_n(t) - f(t))^2 \, dt\right)^{\frac{1}{2}}$$

$$= \sqrt{\frac{1}{h}} \|f_n(t) - f(t)\| \qquad (3.14)$$

From (3.12), we have:

$$\lim_{n \to \infty} \|\mathcal{B}\left\{f_n(t)\right\} - \mathcal{B}\left\{f(t)\right\}\| = 0 \qquad (3.15)$$

Therefore the block pulse operator is a continuous operator.

3. The operator is bounded. Since for linear operators which map between normed linear spaces, the properties of continuity and boundedness are equivalent (Nachbin, 1981). Therefore the block pulse operator is a bounded operator.

3.3   Operation rules of block pulse operators

In this section, we discuss the operation rules of block pulse operator. The discussions are parallel to those about the elementary operations of block pulse series in Section 2.2, but now the emphasis is placed on the asymptotic properties of block pulse series under these operations. Here, the asymptotic approximation between $\mathcal{B}\left\{f(t)\right\}$ and $\mathcal{B}\left\{g(t)\right\}$ means that these vectors satisfy:

$$\lim_{m \to \infty} \|\mathcal{B}\left\{f(t)\right\} - \mathcal{B}\left\{g(t)\right\}\| = 0 \qquad (3.16)$$

under the norm of the space M^m defined in (3.1). For the convenience of expressions, we also denote (3.16) as:

$$\mathcal{B}\left\{f(t)\right\} \sim \mathcal{B}\left\{g(t)\right\} \qquad (3.17)$$

or simply as:

$$F^T \sim G^T \qquad (3.18)$$

1. Constant functions. For a real constant k, we have:

$$\mathcal{B}\left\{k\right\} = kE^T \qquad (3.19)$$

This equation can be derived directly from the definition of block pulse operator.

2. Addition and subtraction of functions. For functions $f(t), g(t) \in L_2[0, T)$, we have:

$$\mathcal{B}\{f(t) \pm g(t)\} = F^T \pm G^T \tag{3.20}$$

This equation can be derived directly from the linearity of block pulse operator.

3. Functions multiplied by a scalar. For a real constant k and a function $f(t) \in L_2[0, T)$, we have:

$$\mathcal{B}\{kf(t)\} = kF^T \tag{3.21}$$

This equation can also be derived directly from the linearity of block pulse operators.

4. Multiplication and division of functions. For functions $f(t), g(t) \in L_2[0, T)$, we have:

$$\mathcal{B}\{f(t)g(t)\} \sim E^T D_F D_G \tag{3.22}$$

If $g(t) \neq 0$, we also have:

$$\mathcal{B}\{f(t)/g(t)\} \sim F^T D_G^{-1} \tag{3.23}$$

In order to prove (3.22), on the one hand, if we set:

$$M = \max_{t \in [0,T)} (|f(t)|, |g(t)|) \tag{3.24}$$

then we obtain:

$$\left\| f(t)g(t) - \hat{f}_m(t)\hat{g}_m(t) \right\| \leq M \left(\left\| f(t) - \hat{f}_m(t) \right\| + \left\| g(t) - \hat{g}_m(t) \right\| \right) \tag{3.25}$$

According to (3.8), we have:

$$\lim_{m \to \infty} \left\| f(t)g(t) - \hat{f}_m(t)\hat{g}_m(t) \right\| = 0 \tag{3.26}$$

The continuity of block pulse operators gives:

$$\lim_{m \to \infty} \left\| \mathcal{B}\{f(t)g(t)\} - \mathcal{B}\left\{\hat{f}_m(t)\hat{g}_m(t)\right\} \right\| = 0 \tag{3.27}$$

or

$$\mathcal{B}\{f(t)g(t)\} \sim \mathcal{B}\left\{\hat{f}_m(t)\hat{g}_m(t)\right\} \tag{3.28}$$

While on the other hand, due to the disjointness of block pulse functions:

$$\begin{aligned} \hat{f}_m(t)\hat{g}_m(t) &= \left(\sum_{i=1}^{m} f_i \phi_i(t) \right) \left(\sum_{j=1}^{m} g_j \phi_j(t) \right) \\ &= \sum_{i=1}^{m} f_i g_i \phi_i(t) \end{aligned} \tag{3.29}$$

the right-hand side of (3.28) becomes:

$$\mathcal{B}\left\{\hat{f}_m(t)\hat{g}_m(t)\right\} = \begin{pmatrix} f_1 g_1 & f_2 g_2 & \cdots & f_m g_m \end{pmatrix} \tag{3.30}$$

Therefore (3.22) is proved.

The proof of (3.23) is similar to the case of multiplication. On the one hand, we can obtain:

$$\lim_{m \to \infty} \left\| f(t)/g(t) - \hat{f}_m(t)/\hat{g}_m(t) \right\| = 0 \tag{3.31}$$

and

$$\mathcal{B}\left\{ f(t)/g(t) \right\} \sim \mathcal{B}\left\{ \hat{f}_m(t)/\hat{g}_m(t) \right\} \tag{3.32}$$

While on the other hand, the disjointness of block pulse functions gives:

$$
\begin{aligned}
\hat{f}(t)/\hat{g}(t) &= \sum_{i=1}^{m} f_i \phi_i(t) \Big/ \sum_{j=1}^{m} g_j \phi_j(t) \\
&= \sum_{i=1}^{m} (f_i/g_i)\phi_i(t)
\end{aligned}
\tag{3.33}
$$

to let the right-hand side of (3.32) become:

$$\mathcal{B}\left\{ \hat{f}_m(t)/\hat{g}_m(t) \right\} = \left(\begin{array}{cccc} f_1/g_1 & f_2/g_2 & \cdots & f_m/g_m \end{array} \right) \tag{3.34}$$

Therefore (3.23) is proved.

5. Integration and derivation of functions. For a function $f(t) \in L_2[0,T]$, we have:

$$\mathcal{B}\left\{ \int_0^t f(t)dt \right\} \sim F^T P \tag{3.35}$$

If $f(t)$ is differentiable, we also have:

$$\mathcal{B}\left\{ \frac{df(t)}{dt} \right\} \sim F^T P^{-1} - f_0^{(0)} E^T P^{-1} \tag{3.36}$$

where P is the integration operational matrix defined in (2.35).

In the proof of (3.35), on the one hand, since $\hat{f}_m(t) \in L_2[0,T]$ satisfies:

$$\lim_{m \to \infty} \hat{f}_m(t) = f(t) \tag{3.37}$$

for all $t \in [0,T]$, we have:

$$\lim_{m \to \infty} \int_0^t \hat{f}_m(t)dt = \int_0^t f(t)dt \tag{3.38}$$

according to the Lebesgue's convergence theorem (Titchmarsh, 1960). Since the functions $f(t)$ and $\hat{f}_m(t)$ in the integrands are bounded and the upper limit of integration $t \in (0,T)$ is finite, the integrals in (3.38) are also bounded. This ensures:

$$\lim_{m \to \infty} \left\| \int_0^t \hat{f}_m(t)dt - \int_0^t f(t)dt \right\| = 0 \tag{3.39}$$

Thus, the continuity of block pulse operator gives:

$$\lim_{m \to \infty} \left\| \mathcal{B}\left\{ \int_0^t \hat{f}_m(t)dt \right\} - \mathcal{B}\left\{ \int_0^t f(t)dt \right\} \right\| = 0 \tag{3.40}$$

or

$$\mathcal{B}\left\{\int_0^t \hat{f}_m(t)dt\right\} \sim \mathcal{B}\left\{\int_0^t f(t)dt\right\} \tag{3.41}$$

On the other hand, if we set:

$$\int_0^t \phi_i(t)dt \sim \sum_{j=1}^m c_{ij}\phi_j(t) \tag{3.42}$$

we can obtain:

$$c_{ij} = \begin{cases} 0 & \text{for } i > j \\ h/2 & \text{for } i = j \\ h & \text{for } i < j \end{cases} \tag{3.43}$$

from (2.29), (2.30), (2.31) and (1.14). According to this relation, (3.42) can be expressed in a vector form:

$$\int_0^t \Phi(t)dt \sim P\Phi(t) \tag{3.44}$$

and the integral of the function $\hat{f}_m(t)$ becomes:

$$\begin{aligned} \int_0^t \hat{f}_m(t)dt &= \int_0^t F^T\Phi(t)dt \\ &\sim F^T P\Phi(t) \end{aligned} \tag{3.45}$$

Thus, the left-hand side of (3.41) becomes:

$$\mathcal{B}\left\{\int_0^t \hat{f}_m(t)dt\right\} \sim F^T P \tag{3.46}$$

and (3.35) is proved.

The proof of (3.36) is rather simple. On the one hand, from

$$\int_0^t \frac{df(t)}{dt}dt = f(t) - f_0^{(0)} \tag{3.47}$$

we obtain:

$$\mathcal{B}\left\{\int_0^t \frac{df(t)}{dt}dt\right\} = \mathcal{B}\left\{f(t)\right\} - f_0^{(0)}E^T \tag{3.48}$$

On the other hand, according to (3.35), we have:

$$\mathcal{B}\left\{\int_0^t \frac{df(t)}{dt}dt\right\} \sim \mathcal{B}\left\{\frac{df(t)}{dt}\right\}P \tag{3.49}$$

Therefore (3.36) is proved.

For multiple integrals and successive derivatives, we have the following rules respectively:

$$\mathcal{B}\left\{\underbrace{\int_0^t \cdots \int_0^t}_{k \text{ times}} f(t)\,dt \cdots dt\right\} \sim F^T P^k \tag{3.50}$$

and

$$B\left\{\frac{d^k f(t)}{dt^k}\right\} \sim F^T P^{-k} - \sum_{i=0}^{k-1} f_0^{(i)} E^T P^{-(k-i)} \tag{3.51}$$

Both of these rules can be proved using the principle of induction.

6. Convolution integral of functions. For functions $f(t), g(t) \in L_2[0, T)$, we have:

$$B\left\{\int_0^t f(\tau)g(t-\tau)d\tau\right\} \sim \frac{h}{2} F^T J_G$$

$$\sim \frac{h}{2} G^T J_F \tag{3.52}$$

where J_G and J_F are the convolution operational matrices defined in (2.62) and (2.63).

In the proof of (3.52), on the one hand, since $\hat{f}_m(\tau)$, $\hat{g}_m(t-\tau) \in L_2[0, T)$ satisfy:

$$\lim_{m\to\infty} \hat{f}_m(\tau)\hat{g}_m(t-\tau) = f(\tau)g(t-\tau) \tag{3.53}$$

for all $t \in [0, T)$, we have:

$$\lim_{m\to\infty} \int_0^t \hat{f}_m(\tau)\hat{g}_m(t-\tau)d\tau = \int_0^t f(\tau)g(t-\tau)d\tau \tag{3.54}$$

according to the Lebesgue's convergence theorem. Since the functions $f(t)$, $g(t)$, $\hat{f}_m(t)$ and $\hat{g}_m(t)$ in the integrands are bounded and the upper limit of integration $t \in (0, T)$ is finite, the integrals in (3.54) are also bounded. This ensures:

$$\lim_{m\to\infty} \left\| \int_0^t \hat{f}_m(\tau)\hat{g}_m(t-\tau)d\tau - \int_0^t f(\tau)g(t-\tau)d\tau \right\| = 0 \tag{3.55}$$

Thus, the continuity of block pulse operator gives:

$$\lim_{m\to\infty} \left\| B\left\{\int_0^t \hat{f}_m(\tau)\hat{g}_m(t-\tau)d\tau\right\} - B\left\{\int_0^t f(\tau)g(t-\tau)d\tau\right\} \right\| = 0 \tag{3.56}$$

or

$$B\left\{\int_0^t \hat{f}_m(\tau)\hat{g}_m(t-\tau)d\tau\right\} \sim B\left\{\int_0^t f(\tau)g(t-\tau)d\tau\right\} \tag{3.57}$$

On the other hand, the derivation from (2.54) to (2.61) gives:

$$\int_0^t \hat{f}_m(\tau)\hat{g}_m(t-\tau)d\tau \sim F^T J_G \Phi(t)$$

$$\sim G^T J_F \Phi(t) \tag{3.58}$$

Therefore the left-hand side of (3.57) becomes:

$$B\left\{\int_0^t \hat{f}_m(\tau)\hat{g}_m(t-\tau)d\tau\right\} \sim F^T J_G$$

$$\sim G^T J_F \tag{3.59}$$

and (3.52) is proved.

7. Functions containing time delay $f(t - \tau)$. For a function $f(t) \in L_2[0, T)$ and $\tau \geq 0$, we have:

$$\mathcal{B}\{f(t - \tau)\} \sim F^T\left((1 - \lambda)H^q + \lambda H^{q+1}\right) \tag{3.60}$$

In the proof of (3.60), on the one hand, since $\hat{f}_m(t - \tau) \in L_2[0, T)$ satisfies:

$$\lim_{m \to \infty} \left\|\hat{f}_m(t - \tau) - f(t - \tau)\right\| = 0 \tag{3.61}$$

we have:

$$\lim_{m \to \infty} \left\|\mathcal{B}\left\{\hat{f}_m(t - \tau)\right\} - \mathcal{B}\{f(t - \tau)\}\right\| = 0 \tag{3.62}$$

or

$$\mathcal{B}\left\{\hat{f}_m(t - \tau)\right\} \sim \mathcal{B}\{f(t - \tau)\} \tag{3.63}$$

according to the continuity of block pulse operator. On the other hand, the result of (2.78) gives:

$$\mathcal{B}\left\{\hat{f}_m(t - \tau)\right\} = F^T\left((1 - \lambda)H^q + \lambda H^{q+1}\right) \tag{3.64}$$

Therefore (3.60) is proved.

Above are the asymptotic properties of block pulse series under some operation rules. Although these properties can also be analysed directly under the expressions of block pulse series, the discussions here under the block pulse operator are much plainer. Using this operator, the operation rules of block pulse functions can also be written in shorter forms, so that expressions of problems can be simplified. To show this, here we solve the same problem of Example 2.1 once more.

Example 3.2 Find the image of the double integral of the function $f(t) = t^2$ under the block pulse operator, in the interval $t \in [0, 1)$ with $m = 4$.

Using the operation rule (3.50) directly, we have:

$$\begin{aligned}
\mathcal{B}\left\{\int_0^t \int_0^t f(t)dtdt\right\} &= \mathcal{B}\{f(t)\}\, P^2 \\
&= \left(\ 0.0208 \quad 0.1458 \quad 0.3958 \quad 0.7708\ \right) P^2 \\
&= \left(\ 0.0003 \quad 0.0036 \quad 0.0179 \quad 0.0589\ \right)
\end{aligned}$$

Obviously, this expression is more compact than the one in Example 2.1. In fact, both block pulse series and block pulse operators can be used in solving problems via block pulse functions, because they are only two different ways of the same essence.

3.4   Extensions of block pulse operators

In the previous sections, the block pulse operator introduced is concerned with scalars. This operator can also be extended to vectors and matrices. The extension is rather simple, because the operator can be applied separately to each entry in the vectors and matrices. Concretely speaking, if $f(t)$ is a $r \times 1$ vector or a $r \times n$ matrix, we define $\mathcal{B}\{f(t)\}$ as a $r \times m$ matrix or a $r \times n \times m$ object respectively, by applying (3.9) to each entry of $f(t)$. In the case where $f(t)$ is a matrix, the object defined is three-dimensional, but it can also be manipulated as an "extended matrix" with the usual operation rules of matrices. For expressing this extended definition mathematically, we introduce a new notation $[[\text{object}]]_k$ which indicates all the entries of this object with the last subscript k. For example, the notation $[[\mathcal{B}\{f(t)\}]]_k$ expresses the shadow parts in Figure 3.1. These shadow parts show that $[[\mathcal{B}\{f(t)\}]]_k$ has the same dimension as $f(t)$, no matter whether $f(t)$ is a scalar, a vector or a matrix.

Using this notation, $\mathcal{B}\{f(t)\}$ can be expressed by a $1 \times m$ vector in all the scalar, vector and matrix cases:

$$\mathcal{B}\{f(t)\} = \left(\ [[\mathcal{B}\{f(t)\}]]_1 \quad [[\mathcal{B}\{f(t)\}]]_2 \quad \cdots \quad [[\mathcal{B}\{f(t)\}]]_m\ \right) \qquad (3.65)$$

Therefore the operation rules of block pulse operator can be described uniformly. Here are some examples in which the dimensions of $f(t), g(t)$ are chosen properly to suit the needs of matrix operations. For a constant vector or matrix K, we have:

$$[[\mathcal{B}\{K\}]]_k = K \qquad (3.66)$$

where $k = 1, 2, \ldots, m$. For the rule of addition, we have:

$$\mathcal{B}\{f(t) + g(t)\} = \mathcal{B}\{f(t)\} + \mathcal{B}\{g(t)\} \qquad (3.67)$$

For the rule of multiplication, we have:

$$[[\mathcal{B}\{f(t)g(t)\}]]_k \sim [[\mathcal{B}\{f(t)\}]]_k[[\mathcal{B}\{g(t)\}]]_k \qquad (3.68)$$

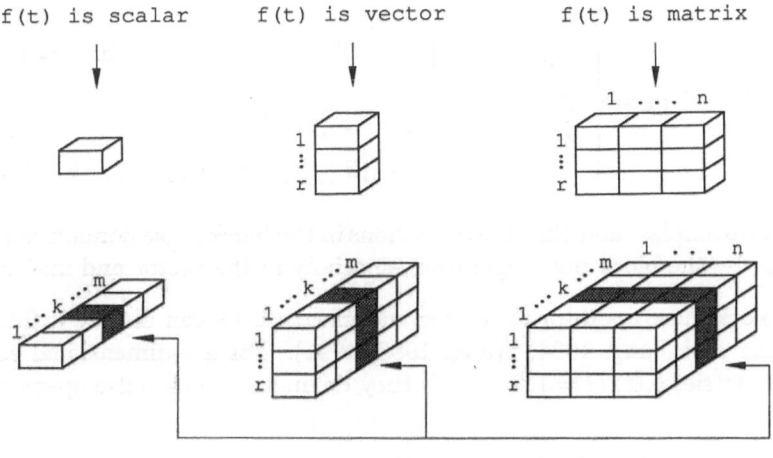

all the elements of $\mathcal{B}\{f(t)\}$
whose last subscript is k

Figure 3.1: Meaning of the notation $[[\mathcal{B}\{f(t)\}]]_k$.

where $k = 1, 2, \ldots, m$. We also have the rule of integration and derivation:

$$\mathcal{B}\left\{\int_0^t f(t)dt\right\} \sim \mathcal{B}\left\{f(t)\right\} P \tag{3.69}$$

and

$$\mathcal{B}\left\{\frac{df(t)}{dt}\right\} \sim \mathcal{B}\left\{f(t)\right\} P^{-1} - (f_0^{(0)} \otimes E^T)P^{-1} \tag{3.70}$$

where $\otimes$ denotes Kronecker product. Since both (3.69) and (3.70) can be written as:

$$[[\mathcal{B}\left\{f(t)\right\} P]]_k = \frac{h}{2}[[\mathcal{B}\left\{f(t)\right\}]]_k + h\sum_{j=1}^{k-1}[[\mathcal{B}\left\{f(t)\right\}]]_j \tag{3.71}$$

and

$$[[\mathcal{B}\left\{f(t)\right\} P^{-1}]]_k = \frac{2}{h}\left((-1)^k f_0^{(0)} + [[\mathcal{B}\left\{f(t)\right\}]]_k + 2\sum_{j=1}^{k-1}(-1)^{k-j}[[\mathcal{B}\left\{f(t)\right\}]]_j\right) \tag{3.72}$$

we obtain the recursive formulas of integration and derivation which are similar to those discussed in Section 2.4:

$$[[\mathcal{B}\left\{f(t)\right\} P]]_k = \begin{cases} \dfrac{h}{2}[[\mathcal{B}\left\{f(t)\right\}]]_1 & \text{for } k = 1 \\[2mm] \dfrac{h}{2}\left([[\mathcal{B}\left\{f(t)\right\}]]_k + [[\mathcal{B}\left\{f(t)\right\}]]_{k-1}\right) & \\[2mm] \qquad +[[\mathcal{B}\left\{f(t)\right\} P]]_{k-1} & \text{for } k = 2, 3, \ldots, m \end{cases} \tag{3.73}$$

and

$$[[\mathcal{B}\left\{f(t)\right\} P^{-1}]]_k = \begin{cases} \dfrac{2}{h}[[\mathcal{B}\left\{f(t)\right\}]]_1 - \dfrac{2}{h}f_0^{(0)} & \text{for } k = 1 \\[2mm] \dfrac{2}{h}\left([[\mathcal{B}\left\{f(t)\right\}]]_k - [[\mathcal{B}\left\{f(t)\right\}]]_{k-1}\right) & \\[2mm] \qquad -[[\mathcal{B}\left\{f(t)\right\} P^{-1}]]_{k-1} & \text{for } k = 2, 3, \ldots, m \end{cases} \tag{3.74}$$

The above examples show that the operations in the block pulse domain can be expressed compactly by the block pulse operator, especially in the vector and matrix cases.

The block pulse operator for vector and matrix cases can also be defined in another way (Wang and Jiang, 1984; Wang, 1983, 1991). For a r-dimensional column vector $Y(t)$ with entries $y_i(t)$ $(i = 1, 2, \ldots, r)$, they define the block pulse operator as a $r \times m$ matrix:

$$\mathcal{B}\left\{Y(t)\right\} = \begin{pmatrix} \mathcal{B}\left\{y_1(t)\right\} \\ \mathcal{B}\left\{y_2(t)\right\} \\ \vdots \\ \mathcal{B}\left\{y_r(t)\right\} \end{pmatrix} \tag{3.75}$$

For an $n \times r$ matrix $A(t)$ with entries $a_{i,j}(t)$ $(i = 1, 2, \ldots, n; j = 1, 2, \ldots, r)$, they define the block pulse operator as an $n \times rm$ matrix:

$$\mathcal{B}\{A(t)\} = \begin{pmatrix} \mathcal{B}\{a_{1,1}(t)\} & \mathcal{B}\{a_{1,2}(t)\} & \cdots & \mathcal{B}\{a_{1,r}(t)\} \\ \mathcal{B}\{a_{2,1}(t)\} & \mathcal{B}\{a_{2,2}(t)\} & \cdots & \mathcal{B}\{a_{2,r}(t)\} \\ \vdots & \vdots & \ddots & \vdots \\ \mathcal{B}\{a_{n,1}(t)\} & \mathcal{B}\{a_{n,2}(t)\} & \cdots & \mathcal{B}\{a_{n,r}(t)\} \end{pmatrix} \tag{3.76}$$

The purpose of these extended definitions is to avoid the three-dimensional objects. For describing the operation rules of these block pulse operators, a $rm \times m$ matrix $\widetilde{D}_Y$, which is constituted by r submatrices D_{Y_i}, is also used:

$$\widetilde{D}_Y = \begin{pmatrix} D_{Y_1} \\ D_{Y_2} \\ \vdots \\ D_{Y_r} \end{pmatrix} . \tag{3.77}$$

where all the submatrices D_{Y_i} $(i = 1, 2, \ldots, r)$ are diagonal matrices determined by $\mathcal{B}\{y_i(t)\}$, respectively.

Here are some operation rules of block pulse operator based on these definitions and notations. For the multiplication, in case of $A(t)y(t)$, we have:

$$\mathcal{B}\{A(t)y(t)\} \sim \mathcal{B}\{A(t)\}\,\widetilde{D}_Y \tag{3.78}$$

and in case of $A(t)y(t)u(t)$ where $u(t)$ is a scalar, we have:

$$\mathcal{B}\{A(t)y(t)u(t)\} \sim \mathcal{B}\{A(t)\}\,\widetilde{D}_Y D_U \tag{3.79}$$

Equations (3.78) and (3.79) show that different formulas should be used for different cases of multiplications. For the integration and derivation of a vector $Y(t)$, the formulas are analogous to (3.69), (3.70), (3.73) and (3.74), only $[[\mathcal{B}\{f(t)\}\,P]]_k$ and $[[\mathcal{B}\{f(t)\}]]_k$ should be replaced by the kth column of the $r \times m$ matrices $\mathcal{B}\{Y(t)\}\,P$ and $\mathcal{B}\{Y(t)\}$, respectively. Details about block pulse operators and their applications can be found in the papers of Wang and Jiang, (1984, 1985a,b), Wang (1983a,b, 1991), Zhu and Lu, (1987, 1988b).

3.5   Notes on block pulse operators

In control theory and systems science, the Laplace transform is a useful tool. A lot of problems, especially those of time-invariant linear systems, can first be transformed into much simpler forms in the Laplace domain, and then they can be solved more easily and efficiently.

If we compare some operation rules of the block pulse operator with those of the Laplace transform, we can notice that their forms are somewhat similar. In Table 3.1,

Time domain	Block pulse domain	Laplace domain
$f(t)$	F^T	$F(s)$
$f(t) \pm g(t)$	$F^T \pm G^T$	$F(s) \pm G(s)$
$kf(t)$	kF^T	$kF(s)$
$f(t)g(t)$	$E^T D_F D_G$	$\dfrac{1}{2\pi j} F(s) * G(s)$
$\displaystyle\int_0^t f(t)dt$	$F^T P$	$\dfrac{1}{s} F(s)$
$\underbrace{\displaystyle\int_0^t \cdots \int_0^t f(t)dt \cdots dt}_{k \text{ times}}$	$F^T P^k$	$\dfrac{1}{s^k} F(s)$
$\dfrac{df(t)}{dt}$	$F^T P^{-1} - f_0^{(0)} E^T P^{-1}$	$sF(s) - f_0^{(0)}$
$\dfrac{d^k f(t)}{dt^k}$	$F^T P^{-k} - E^T \displaystyle\sum_{i=0}^{k-1} f_0^{(i)} P^{-(k-i)}$	$s^k F(s) - \displaystyle\sum_{i=0}^{k-1} f_0^{(i)} s^{k-i-1}$
$f(t) * g(t)$	$F^T J_G \quad$ or $\quad G^T J_F$	$F(s)G(s)$
$f(t - \tau)$	$F^T \left((1-\lambda)H^q + \lambda H^{q+1} \right)$	$e^{-\tau s} F(s)$

Table 3.1: Some operation rules in the block pulse and Laplace domain.

we list some operations both in the block pulse domain and in the Laplace domain to show their similarities. Since the Laplace transform is applied frequently in control and systems science and is familiar to the people, many expressions in the block pulse domain can be directly adopted with little modifications from the results obtained in the Laplace domain. The block diagram of systems is a typical example which will be discussed further in Chapter 6.

In Table 3.1, we can notice that the matrices P and P^{-1} in the block pulse domain play the same roles as s^{-1} and s in the Laplace domain, respectively. Since s^{-1} and s are the integral and derivative operators in the Laplace domain, the relations:

$$P \longleftrightarrow \frac{1}{s} \tag{3.80}$$

and

$$P^{-1} \longleftrightarrow s \tag{3.81}$$

are extremely helpful when problems containing integrals and derivatives are studied in the block pulse domain.

Since block pulse functions have the special property of disjointness, some formulas obtained in the block pulse domain are simpler than those obtained in the Laplace domain. As an example, for the case of function multiplications in the time domain, the block pulse operator leads it to the multiplication of matrices in the block pulse domain, but the Laplace transform leads it to the convolution integral of functions in the Laplace domain. Due to this simpler formula about function multiplications, certain problems related to time-varying linear systems can be dealt with more easily by block pulse functions.

Although some formulas in the block pulse domain and in the Laplace domain have similarities in the form, their meanings are essentially different. As one difference, the Laplace transform is implemented in the interval $t \in [0, \infty)$, whereas the block pulse operator is defined only in the interval $t \in [0, T)$. As another difference, the results obtained from the Laplace transform are exact, whereas the block pulse operator gives only piecewise constant approximate results with a finite number m in the practical uses.

About the approximate results which are obtained from the block pulse operator, one thing should also be mentioned. Although the asymptotic property of the approximations guarantees that the mean square errors between the analytical solutions and their block pulse series approximations defined in (1.11) can become arbitrary small when the number m approaches infinity, the errors under a finite number m must not arrive to their minimal values after some operations are applied to the image of the block pulse operator. This is the same as the discussion in Example 2.1.

Chapter 4

Block pulse transforms

Although the block pulse series can be applied to approximate original functions which are essentially continuous or piecewise continuous in the whole definition interval, they are somewhat like the discrete sampled signals because the block pulse series take only one value in each subinterval. From this point of view, the method of expressing discrete signals in sampled-data systems, such as the delay operator, can also be adopted for the block pulse series expansions. In this chapter, we will use such methods in the block pulse domain to improve algorithms and simplify computations.

4.1   Definition of block pulse transforms

In Section 2.5, we mentioned that there is a close relation between the Tustin integrator and the block pulse operational matrix of integration. In fact, if we study the constructions of the integration operational matrix P, the one step delay operational matrix H and the unit matrix I together, we can find:

$$P = \frac{h}{2}(I + H)(I - H)^{-1} \tag{4.1}$$

The form of this equation is very similar to that of the Tustin integrator:

$$s^{-1} = \frac{h}{2}\frac{1 + z^{-1}}{1 - z^{-1}} \tag{4.2}$$

where the one step delay operator is $z^{-1} = e^{-hs}$. This close relation can also be expressed in another form by the delay operators z^{-1} and H as follows.

For a power series related to the scalar z^{-1}:

$$x(z^{-1}) = \alpha_0 + \alpha_1 z^{-1} + \alpha_2 z^{-2} + \cdots \tag{4.3}$$

the corresponding power series related to the matrix H is:

$$x(H) = \alpha_0 I + \alpha_1 H + \alpha_2 H^2 + \cdots \tag{4.4}$$

We can notice that the matrix $x(H)$ has a regular upper triangular form, i.e. its first row is composed of the coefficients of the first m terms in the power series $x(z^{-1})$, and its kth row can be obtained by shifting the first row $(k-1)$ positions to the right. Based on this matrix form, we denote the relation between the power series $x(z^{-1})$ and the matrix $x(H)$ as:

$$\alpha_0 + \alpha_1 z^{-1} + \alpha_2 z^{-2} + \cdots \Longleftrightarrow \begin{pmatrix} \alpha_0 & \alpha_1 & \alpha_2 & \cdots & \alpha_{m-1} \\ 0 & \alpha_0 & \alpha_1 & \cdots & \alpha_{m-2} \\ 0 & 0 & \alpha_0 & \cdots & \alpha_{m-3} \\ \vdots & \vdots & \vdots & \ddots & \vdots \\ 0 & 0 & 0 & \cdots & \alpha_0 \end{pmatrix} \qquad (4.5)$$

If the upper triangular matrix in (4.5) is premultiplied by a vector with all entries ones, we obtain a vector:

$$E^T x(H) = \begin{pmatrix} \alpha_0 & \alpha_0 + \alpha_1 & \cdots & \alpha_0 + \cdots + \alpha_{m-1} \end{pmatrix} \qquad (4.6)$$

Since the entries in this vector are the coefficients of the first m terms in the power series $x(z^{-1})/(1 - z^{-1})$ respectively:

$$\frac{x(z^{-1})}{1 - z^{-1}} = \alpha_0 + (\alpha_0 + \alpha_1) z^{-1} + \ldots + (\alpha_0 + \cdots + \alpha_{m-1}) z^{-(m-1)} + \ldots \qquad (4.7)$$

we denote the relation between the power series $x(z^{-1})/(1 - z^{-1})$ and the vector $E^T x(H)$ as:

$$\alpha_0 + (\alpha_0 + \alpha_1) z^{-1} + \ldots + (\alpha_0 + \cdots + \alpha_{m-1}) z^{-(m-1)} + \ldots$$
$$\Longleftrightarrow \begin{pmatrix} \alpha_0 & \alpha_0 + \alpha_1 & \cdots & \alpha_0 + \cdots + \alpha_{m-1} \end{pmatrix} \qquad (4.8)$$

Equations (4.5) and (4.8) can be used in introducing the block pulse transform.

Consider the problem about the inverse Laplace transform of a rational transfer function:

$$F(s) = \frac{b_{n-1} s^{n-1} + b_{n-2} s^{n-2} + \cdots + b_0}{s^n + a_{n-1} s^{n-1} + \cdots + a_0} \qquad (4.9)$$

This problem can be interpreted as the response of a system with the transfer function $sF(s)$ under a unit step excitation $U(s) = 1/s$:

$$\begin{aligned} F(s) &= sF(s) \times \frac{1}{s} \\ &= F_1(s^{-1}) U(s) \\ &= \frac{b_{n-1} + b_{n-2} s^{-1} + \ldots + b_0 s^{-(n-1)}}{1 + a_{n-1} s^{-1} + \ldots + a_0 s^{-n}} U(s) \end{aligned} \qquad (4.10)$$

Here, instead of expressing the transfer function $sF(s)$ in the terms s^k ($k = 1, 2, \ldots, n$), we write it in the terms s^{-k}. The reason of using such expression will be mentioned in Section 4.3. After rewriting (4.10) in the form:

$$\left(1 + a_{n-1} s^{-1} + \ldots + a_0 s^{-n}\right) F(s) = \left(b_{n-1} + b_{n-2} s^{-1} + \ldots + b_0 s^{-(n-1)}\right) U(s) \qquad (4.11)$$

and replacing each power of s^{-1} of the above equation with the respective power of P, the block pulse coefficients of the inverse Laplace transform of $F(s)$ can be obtained as:

$$F^T = E^T \left(b_{n-1}I + b_{n-2}P + \ldots + b_0 P^{n-1} \right) \left(I + a_{n-1}P + \ldots + a_0 P^n \right)^{-1} \qquad (4.12)$$

In the right-hand side of (4.12), the vector E has all entries ones which represents the block pulse coefficient vector of the unit step excitation, and the remaining part is an upper triangular matrix because the power and inverse of upper triangular matrices are still upper triangular matrices.

For expressing a discrete signal $f^*(t)$ in sampled-data systems, we can use the z-transform (Franklin, Powell and Workman, 1990):

$$F(z) = \sum_{i=0}^{\infty} f^*(ih) z^{-i} \qquad (4.13)$$

where h is the sampling period, and the coefficients of the terms z^{-i} $(i = 0, 1, \ldots)$ in the power series are the values of the sampled signal $f^*(t)$ at the corresponding time instants $t = ih$. Now we adopt this method of expressing discrete sampled signals and apply it to express the coefficients of block pulse series. Noticing that in (4.12), the vector E^T is related to:

$$E^T \Longleftrightarrow 1 + z^{-1} + z^{-2} + \cdots = \frac{1}{1 - z^{-1}} \qquad (4.14)$$

and the upper triangular matrix part is related to:

$$\left(b_{n-1}I + b_{n-2}P + \ldots + b_0 P^{n-1} \right) \left(I + a_{n-1}P + \ldots + a_0 P^n \right)^{-1}$$
$$\Longleftrightarrow F_1 \left(\frac{h}{2} \frac{1 + z^{-1}}{1 - z^{-1}} \right) \qquad (4.15)$$

we can obtain the block pulse coefficients in the vector F^T from the coefficients of the first m terms in the power series:

$$\frac{1}{1 - z^{-1}} F_1 \left(\frac{h}{2} \frac{1 + z^{-1}}{1 - z^{-1}} \right) = f_0 + f_1 z^{-1} + \cdots + f_{m-1} z^{-(m-1)} + \cdots \qquad (4.16)$$

according to (4.8). Therefore we define the block pulse transform with respect to the Laplace transform of $F(s)$ as:

$$F^s(z) = \frac{1}{1 - z^{-1}} F_1 \left(\frac{h}{2} \frac{1 + z^{-1}}{1 - z^{-1}} \right) \qquad (4.17)$$

In this definition, we use the notation $F^s(z)$ in order to distinguish the block pulse transform from the z-transform $F(z)$ of the sampled signal $f^*(t)$. For the convenience of use, some expressions of block pulse transform are listed in Table 4.1. In fact, the block pulse transform can be applied to functions with general forms, although it is introduced here from the inverse Laplace transform of a rational transfer function (Marszalek, 1984c). Here is the example of an irrational function.

Time function	Block pulse transform
1	$\dfrac{1}{1 - z^{-1}}$
t	$\dfrac{h(1 + z^{-1})}{2(1 - z^{-1})^2}$
$t^n \quad (n=1,2,\ldots)$	$\dfrac{h^n n!(1 + z^{-1})^n}{2^n(1 - z^{-1})^{n+1}}$
e^{-at}	$\dfrac{2}{(ah - 2)z^{-1} + (ah + 2)} \qquad (ah \neq -2)$
te^{-at}	$\dfrac{2h(1 + z^{-1})}{[(ah - 2)z^{-1} + (ah + 2)]^2} \qquad (ah \neq -2)$
$t^n e^{-at} \quad (n=1,2,\ldots)$	$\dfrac{2h^n n!(1 + z^{-1})^n}{[(ah - 2)z^{-1} + (ah + 2)]^{n+1}} \qquad (ah \neq -2)$
$\sin at$	$\dfrac{2ah(1 + z^{-1})}{(a^2h^2 + 4)z^{-2} + 2(a^2h^2 - 4)z^{-1} + (a^2h^2 + 4)}$
$\cos at$	$\dfrac{4(1 - z^{-1})}{(a^2h^2 + 4)z^{-2} + 2(a^2h^2 - 4)z^{-1} + (a^2h^2 + 4)}$

Table 4.1: Some time functions and their block pulse transforms.

Example 4.1 Find the block pulse coefficients of the inverse Laplace transform of an irrational transfer function:

$$F(s) = \frac{1}{\sqrt{s^2 + 1}}$$

using the definition of the block pulse transform, in the interval $t \in [0, 8)$ with $h = 0.25$.

To obtain the requested block pulse coefficients, we first express $sF(s)$ in the form of $F_1(s^{-1})$:

$$sF(s) = \frac{1}{\sqrt{s^{-2} + 1}}$$

and then expand $F^s(z)$ into its power series:

$$
\begin{aligned}
F^s(z) &= \frac{1}{(1 - z^{-1})} \frac{1}{\sqrt{\left(\dfrac{h(1 + z^{-1})}{2(1 - z^{-1})}\right)^2 + 1}} \\
&= 0.9923 + 0.9617z^{-1} + 0.9021z^{-2} + 0.8161z^{-3} + \cdots
\end{aligned}
$$

The coefficients in the above power series are the requested block pulse coefficients of $f(t)$. The analytical solution of this inverse Laplace transform problem is the Bessel

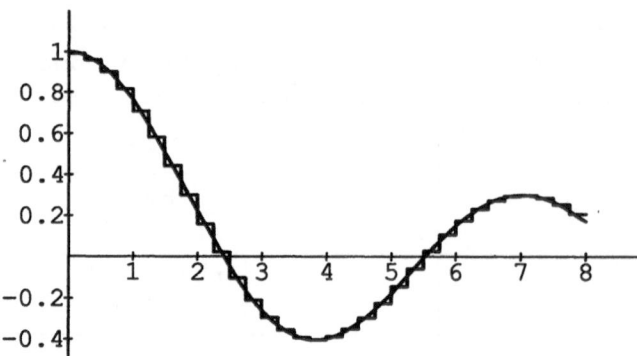

Figure 4.1: Block pulse series approximation of the Bessel function $J_0(t)$.

function of the first kind $J_0(t)$. This exact solution and its block pulse series approximation are illustrated in Figure 4.1.

4.2   Operation rules of block pulse transforms

Operation rules of the block pulse transform can be discussed based on its definition (4.17), but they are similar to the elementary operations in Section 2.2. Therefore we mention them only briefly in this section. For simplicity of expressions in the operation rules below, sometimes we also write the block pulse transform of functions in power series forms like:

$$F^s(z) = \sum_{i=0}^{\infty} f_i z^{-i} \tag{4.18}$$

instead of in function forms like (4.17).

1. **Constant functions.** For $x(t) = k$, we have:

$$X^s(z) = \frac{k}{1 - z^{-1}} \tag{4.19}$$

2. **Addition and subtraction of functions.** For $x(t) = f(t) \pm g(t)$, we have:

$$X^s(z) = F^s(z) \pm G^s(z) \tag{4.20}$$

3. **Functions multiplied by a scalar.** For $x(t) = kf(t)$, we have:

$$X^s(z) = kF^s(z) \tag{4.21}$$

4. **Multiplication and division of functions.** For $x(t) = f(t)g(t)$, we have:

$$X^s(z) = \sum_{i=0}^{\infty} f_i g_i z^{-i} \tag{4.22}$$

and for $x(t) = f(t)/g(t)$ with $g(t) \neq 0$, we have:

$$X^s(z) = \sum_{i=0}^{\infty}(f_i/g_i)z^{-i} \tag{4.23}$$

5. Integration and derivation of functions. For $x(t) = \int_0^t f(t)dt$, we have:

$$X^s(z) = \frac{h}{2}\frac{1+z^{-1}}{1-z^{-1}}F^s(z) \tag{4.24}$$

and for $x(t) = \dfrac{df(t)}{dt}$, we have:

$$X^s(z) = \frac{2}{h}\frac{1-z^{-1}}{1+z^{-1}}F^s(z) - f_0^{(0)}\frac{2}{h}\frac{1}{1+z^{-1}} \tag{4.25}$$

where $f_0^{(0)}$ is the initial value of $f(t)$.

6. Convolution integral of functions. For $x(t) = \int_0^t f(\tau)g(t-\tau)d\tau$, we have:

$$X^s(z) = \frac{h}{2}(1+z^{-1})F^s(z)G^s(z) \tag{4.26}$$

7. Functions containing time delay $f(t-\tau)$. For $x(t) = f(t-\tau)$, we have:

$$X^s(z) = z^{-q}F^s(z) \tag{4.27}$$

where $\tau = qh$ and q is a positive integer.

Although the above operation rules are similar to those of block pulse series discussed in the previous chapters, they can only be applied in the cases where the Laplace transform of functions are obtained, because the block pulse transform is originally defined from the Laplace transform. This is one restriction of the block pulse transform method. Moreover, (4.17) indicates that the block pulse transform is defined from the approximation of a step response. To obtain the block pulse coefficients of a function, more approximations are introduced by the block pulse transform than by the formula (1.14), because the former may introduce errors in several steps, whereas the latter can give the minimal mean square errors between the function and its approximate block pulse series. Since the block pulse coefficients obtained from the block pulse transform are worse than or at most the same as the exact block pulse coefficients, the operation rules applied on these data can only give results worse than or at most the same as those obtained from the block pulse series expansion. This is also one of the deficiencies of the block pulse transform method. Here is a simple example.

Example 4.2 Find the first four block pulse coefficients of the double integral of the function $f(t) = t^2$ with $h = 0.25$, using the integral operation rule of the block pulse transform.

We denote $g(t)$ as the double integral of $f(t)$. Since the block pulse transform of $f(t)$ is:

$$F^s(z) = \frac{2}{1 - z^{-1}} \left(\frac{h}{2} \frac{1 + z^{-1}}{1 - z^{-1}} \right)^2$$

we obtain the block pulse transform of the result through applying the operation rule (4.24) twice:

$$
\begin{aligned}
G^s(z) &= \frac{2}{1 - z^{-1}} \left(\frac{h}{2} \frac{1 + z^{-1}}{1 - z^{-1}} \right)^4 \\
&= 0.0005 + 0.0044 z^{-1} + 0.0200 z^{-2} + 0.0630 z^{-3}
\end{aligned}
$$

The same problem was solved in Example 2.1 through the block pulse series expansion. To compare the accuracies of the approximate results of the two methods, we use the sum of the square error between the exact and approximate block pulse coefficients of the result as a criterion:

$$e_j = \sum_{k=1}^{m} (g_k - g_{j,k})^2 \tag{4.28}$$

In this criterion, the approximate block pulse coefficients of the results are denoted as $g_{j,k}$ $(k = 1, 2, \ldots, m)$ with the subscripts $j = 1$ for Example 2.1 and $j = 2$ for Example 4.2, respectively. The exact block pulse coefficients of the result obtained from (1.14) are denoted as g_k. For these two methods, the criterion gives the values $e_1 = 0.000085$ and $e_2 = 0.000193$ which show that the result obtained from the block pulse transform is not so good as the one obtained from the block pulse series expansion.

As mentioned above, we can further point out the reason of the worse result in this example. For the case $f(t) = t^2$, (4.12) becomes:

$$F^T = E^T \left(2P^2 \right)$$

It means that the block pulse coefficients of the function t^2 is obtained from a step function with a double integral. Since the integration operational matrix P is only an approximation of the integral operation, the block pulse coefficients of $f(t)$ in Example 4.2 are surely worse than those in Example 2.1, and this makes the final result worse.

4.3   Extensions of block pulse transforms

In digital simulation of physical systems, some numerical methods are oriented to the discrete integrators, e.g. Tustin integrator (Tustin, 1947), Madwed integrators (Madwed, 1950), Boxer-Thaler integrators (Boxer and Thaler, 1956), Halijak integrators (Halijak, 1960) and so on. Among these operators, the Tustin integrator is of great utility because it is quite accurate and does not introduce spurious solutions. Since the orders of integrals are related by the corresponding powers, the Tustin integrator is ideally suited for operational calculus applications, but these powers accumulate also

Laplace integrators	Madwed integrators
s^{-1}	$\dfrac{h}{2}\dfrac{1+z^{-1}}{1-z^{-1}}$
s^{-2}	$\dfrac{h^2}{6}\dfrac{1+4z^{-1}+z^{-2}}{(1-z^{-1})^2}$
s^{-3}	$\dfrac{h^3}{24}\dfrac{1+11z^{-1}+11z^{-2}+z^{-3}}{(1-z^{-1})^3}$
s^{-4}	$\dfrac{h^4}{120}\dfrac{1+26z^{-1}+66z^{-2}+26z^{-3}+z^{-4}}{(1-z^{-1})^4}$

Laplace integrators	Boxer-Thaler integrators
s^{-1}	$\dfrac{h}{2}\dfrac{1+z^{-1}}{1-z^{-1}}$
s^{-2}	$\dfrac{h^2}{12}\dfrac{1+10z^{-1}+z^{-2}}{(1-z^{-1})^2}$
s^{-3}	$\dfrac{h^3}{2}\dfrac{z^{-1}+z^{-2}}{(1-z^{-1})^3}$
s^{-4}	$\dfrac{h^4}{6}\dfrac{z^{-1}+4z^{-2}+z^{-3}}{(1-z^{-1})^4}-\dfrac{h^4}{720}$

Table 4.2: First four Madwed and Boxer-Thaler integrators.

approximate errors at each stage of integral. For multiple integrals, some other discrete integrators, such as Madwed integrators and Boxer-Thaler integrators, are more preferable. The improvements of accuracies under these discrete integrators are evident in the simulation of various models (Rosko, 1972). Here in Table 4.2, we list the first four Madwed and Boxer-Thaler integrators as an example. Unlike the Tustin integrator, these integrators with higher order are not related to those with lower order.

In Section 4.1, we applied the Tustin integrator to the block pulse transform. Because of the same reason as mentioned above, the accumulated errors are as well introduced to the block pulse transform. In order to improve the accuracies of the block pulse transform, we can also use other discrete integrators. Here as examples, we use the Madwed and Boxer-Thaler integrators to realize the extensions of the block pulse transform and to show the improvements of accuracies. In fact, the extensions are rather simple. Since the main thing in introducing the block pulse transform based on the Tustin integrator is to replace each power of s^{-1} in the expression $F_1(s^{-1})$ with the respective power of the Tustin integrator, now the factors s^{-k} in the expression $F_1(s^{-1})$ should only be replaced by the kth-order Madwed or Boxer-Thaler integrator, respectively. But in both extensions, it is not so easy to give a uniform formula for the block pulse transform like (4.17) of the Tustin integrator case, because different order Madwed and Boxer-Thaler integrators have different expressions. To distinguish the block pulse transforms obtained from various discrete integrators as discussed above, we call them the block pulse transforms based on Tustin, Madwed and Boxer-Thaler integrators, respectively.

Example 4.3 Evaluate the block pulse coefficients of the impulse response of a linear system $F(s) = 1/(s^2 + 3s + 2)$ in the interval $t \in [0,5)$ with $m = 20$, and compare the accuracies of the results obtained from the block pulse transforms based on Tustin, Madwed and Boxer-Thaler integrators.

From the transfer function of the system, we first express $sF(s)$ in the form of $F_1(s^{-1})$:

$$sF(s) = \frac{s^{-1}}{1 + 3s^{-1} + 2s^{-2}} \tag{4.29}$$

If we substitute the Tustin integrator for the factor s^{-1} in (4.29), we obtain the block pulse transform:

$$
\begin{aligned}
F^s(z) &= \frac{1}{1-z^{-1}} \frac{\dfrac{h}{2}\dfrac{1+z^{-1}}{1-z^{-1}}}{1 + 3\dfrac{h}{2}\dfrac{1+z^{-1}}{1-z^{-1}} + 2\left(\dfrac{h}{2}\dfrac{1+z^{-1}}{1-z^{-1}}\right)^2} \\
&= 0.0889 + 0.2114z^{-1} + 0.2497z^{-2} + 0.2454z^{-3} + \cdots
\end{aligned} \tag{4.30}
$$

If we substitute the first- and second-order Madwed integrators for the factors s^{-1} and s^{-2} in (4.29) respectively, we obtain the block pulse transform:

$$
\begin{aligned}
F^s(z) &= \frac{1}{1-z^{-1}} \frac{\dfrac{h}{2}\dfrac{1+z^{-1}}{1-z^{-1}}}{1 + 3\dfrac{h}{2}\dfrac{1+z^{-1}}{1-z^{-1}} + 2\dfrac{h^2}{6}\dfrac{1+4z^{-1}+z^{-2}}{(1-z^{-1})^2}} \\
&= 0.0896 + 0.2125z^{-1} + 0.2504z^{-2} + 0.2455z^{-3} + \cdots
\end{aligned} \tag{4.31}
$$

If we substitute the first- and second-order Boxer-Thaler integrators for the factors s^{-1} and s^{-2} in (4.29) respectively, we obtain the block pulse transform:

$$
\begin{aligned}
F^s(z) &= \frac{1}{1-z^{-1}} \frac{\dfrac{h}{2}\dfrac{1+z^{-1}}{1-z^{-1}}}{1 + 3\dfrac{h}{2}\dfrac{1+z^{-1}}{1-z^{-1}} + 2\dfrac{h^2}{12}\dfrac{1+10z^{-1}+z^{-2}}{(1-z^{-1})^2}} \\
&= 0.0902 + 0.2137z^{-1} + 0.2510z^{-2} + 0.2455z^{-3} + \cdots
\end{aligned} \tag{4.32}
$$

The power series in (4.30), (4.31) and (4.32) contain infinite terms, but only the first m coefficients related to the terms z^{-i} $(i = 0, 1, \ldots, m - 1)$ are the requested block pulse coefficients of the impulse response in our problem. To compare the accuracies of the results, we use the criterion same as (4.28), which indicates the sum of the square error between the exact and approximate block pulse coefficients of the results. With the subscripts $j = 1$ fot Tustin, $j = 2$ for Madwed and $j = 3$ for Boxer-Thaler variations respectively, the criterion gives the values $e_1 = 0.000110$, $e_2 = 0.000097$ and $e_3 = 0.000093$, which show that the block pulse transforms based on the Madwed and Boxer-Thaler integrators are superior to the one based on the Tustin integrator.

In Section 4.1, we emphasized that the expression $sF(s)$ should first be rewritten to the form $F_1(s^{-1})$, and then each power of s^{-1} in $F_1(s^{-1})$ is replaced by the respective power of the Tustin integrator. In fact, if we introduce the block pulse transform only

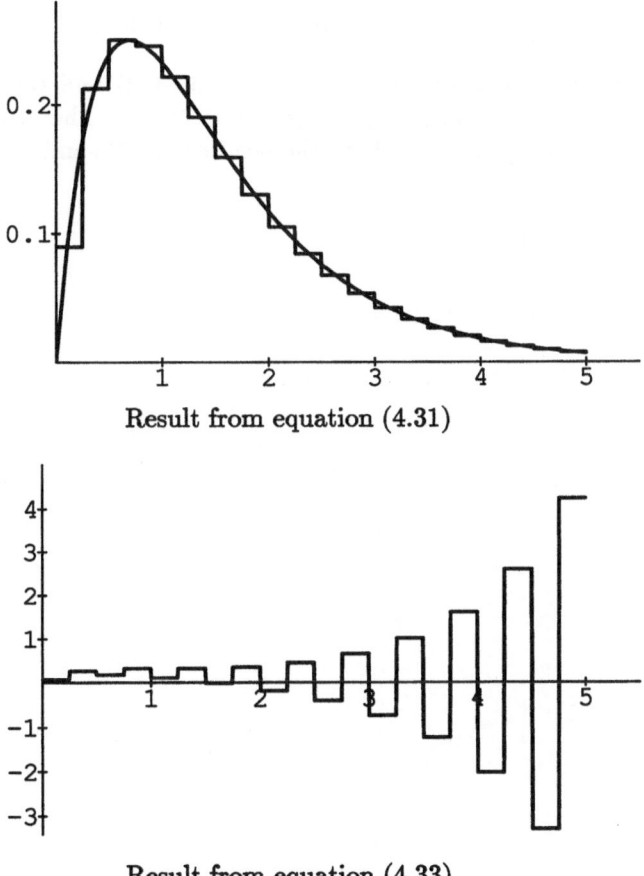

Result from equation (4.31)

Result from equation (4.33)

Figure 4.2: Right and wrong manipulations in block pulse transform.

based on the Tustin integrator, this emphasis is not necessary, because the expression:

$$\cfrac{1}{1-z^{-1}}\cfrac{\dfrac{2}{h}\dfrac{1-z^{-1}}{1+z^{-1}}}{\left(\dfrac{2}{h}\dfrac{1-z^{-1}}{1+z^{-1}}\right)^2+3\dfrac{2}{h}\dfrac{1-z^{-1}}{1+z^{-1}}+2}$$

which is directly obtained from $sF(s) = s/(s^2 + 3s + 2)$ and the reciprocal of Tustin integrator, is equivalent to the one in (4.30). But in the extensions based on the Madwed and Boxer-Thaler integrators, no such equivalence exists. For example, the expression:

$$\cfrac{1}{1-z^{-1}}\cfrac{\dfrac{2}{h}\dfrac{1-z^{-1}}{1+z^{-1}}}{\dfrac{6}{h^2}\dfrac{(1-z^{-1})^2}{1+4z^{-1}+z^{-2}}+3\dfrac{2}{h}\dfrac{1-z^{-1}}{1+z^{-1}}+2}$$

$$= \quad 0.0656 + 0.2698z^{-1} + 0.1815z^{-2} + 0.3305z^{-3} + \cdots \qquad (4.33)$$

which is directly obtained from $sF(s) = s/(s^2 + 3s + 2)$ and the reciprocal of the Madwed integrators, is not equivalent to the one in (4.31). This wrong manipulation brings an uncorrect divergent series. Both the results of (4.31) and (4.33) are illustrated in Figure 4.2.

Chapter 5

Block pulse operational matrices for integrations

In applying the block pulse function technique to control and systems science, the integral operation rules of block pulse series play important roles. This is because differential equations are always involved in the representations of continuous-time models of dynamic systems, and differential operations are always approximated by the corresponding block pulse series through integration operational matrices. In order to apply the block pulse function technique to the problems of continuous-time dynamic systems more efficiently, it is necessary to study further the integral operation rules of block pulse series. In this chapter, we will improve these integral operation rules and introduce the related integration operational matrices.

5.1 Improved integration operational matrix

As an elementary operation rule of block pulse series, the integral of a function can be transformed approximately into the algebraic operation:

$$
\begin{aligned}
g(t) &= \int_0^t f(t)dt \\
&\doteq F^T P \Phi(t) \\
&= \begin{pmatrix} \tilde{g}_1 & \tilde{g}_2 & \cdots & \tilde{g}_m \end{pmatrix} \Phi(t)
\end{aligned}
\tag{5.1}
$$

Here, we denote the block pulse coefficients of $g(t)$ as $\tilde{g}_i$ $(i = 1, 2, \ldots, m)$ to emphasize that they are not the exact block pulse coefficients of $g(t)$ obtained from the formula (1.14), but are the approximations obtained from the operation $F^T P$. In order to reveal the geometrical meaning of these block pulse coefficients, we express them more clearly according to the regular upper triangular form of the integration operational matrix P:

$$
\tilde{g}_i = h(f_1 + f_2 + \ldots + f_{i-1}) + \frac{h}{2} f_i
\tag{5.2}
$$

Using (1.14), we can also rewrite this equation as:

$$
\begin{aligned}
\tilde{g}_i &= \int_0^{(i-1)h} f(t)dt + \frac{1}{2}\int_{(i-1)h}^{ih} f(t)dt \\
&= g((i-1)h) + \frac{g(ih) - g((i-1)h)}{2} \\
&= \frac{g((i-1)h) + g(ih)}{2}
\end{aligned}
\tag{5.3}
$$

It shows that the ith block pulse coefficient obtained from (5.1) is only the mean value of the function $g(t)$ at two points $t_1 = (i-1)h$ and $t_2 = ih$. In other words, the block pulse coefficient $\tilde{g}_i$ is related to the area in the subinterval $t \in [(i-1)h, ih)$ under a linear approximation of $g(t)$. But according to (1.14), the ith block pulse coefficient should be related to the area in the subinterval $t \in [(i-1)h, ih)$ under the curve of $g(t)$. It is easy to see that the errors contained in the block pulse coefficients $\tilde{g}_i$ $(i = 1, 2, \ldots, m)$ are introduced by the inaccurate evaluation of the areas in the subintervals.

In order to improve the accuracy of block pulse coefficients obtained from the integral operation rule, we can first establish some better approximations of $g(t)$, and then evaluate the block pulse coefficients from the areas in the subintervals under these better approximations (Chen and Chung, 1987). Here as an example, we first use the Lagrange's interpolation formula, which passes through three points $t_0 = (i-2)h$, $t_1 = (i-1)h$ and $t_2 = ih$, to approximate $g(t)$ in the subinterval $t \in [(i-1)h, ih)$:

$$
\begin{aligned}
\bar{g}(t) &= g((i-2)h)\frac{(t-(i-1)h)(t-ih)}{2h^2} \\
&- g((i-1)h)\frac{(t-(i-2)h)(t-ih)}{h^2} \\
&+ g(ih)\frac{(t-(i-2)h)(t-(i-1)h)}{2h^2}
\end{aligned}
\tag{5.4}
$$

and then evaluate the ith block pulse coefficient of $g(t)$ from this approximation $\bar{g}(t)$ as an improvement:

$$
\begin{aligned}
\bar{g}_i &= \frac{1}{h}\int_{(i-1)h}^{ih} \bar{g}(t)dt \\
&= -\frac{1}{12}g((i-2)h) + \frac{8}{12}g((i-1)h) + \frac{5}{12}g(ih)
\end{aligned}
\tag{5.5}
$$

Since the function $g(t)$ is the integral of $f(t)$:

$$
\begin{aligned}
g(ih) &= \int_0^{ih} f(t)dt \\
&= h\left(\frac{1}{h}\int_0^h f(t)dt + \frac{1}{h}\int_h^{2h} f(t)dt + \ldots + \frac{1}{h}\int_{(i-1)h}^{ih} f(t)dt\right) \\
&= h(f_1 + f_2 + \ldots + f_i)
\end{aligned}
\tag{5.6}
$$

this relation and the similar ones:

$$
g((i-1)h) = h(f_1 + f_2 + \ldots + f_{i-1})
\tag{5.7}
$$

$$g((i-2)h) = h(f_1 + f_2 + \ldots + f_{i-2}) \qquad (5.8)$$

can be substituted into (5.5) to obtain:

$$\bar{g}_i = h\left(f_1 + f_2 + \ldots + f_{i-2} + \frac{13}{12}f_{i-1} + \frac{5}{12}f_i\right) \qquad (5.9)$$

But this improvement can only be applied to the cases $i = 3, 4, \ldots, m$ because f_{-1} and f_0 do not exist. For the case $i = 2$, the equation (5.5) gives directly the improvement:

$$
\begin{aligned}
\bar{g}_2 &= -\frac{1}{12}g(0) + \frac{8}{12}g(h) + \frac{5}{12}g(2h) \\
&= h\left(\frac{13}{12}f_1 + \frac{5}{12}f_2\right) \qquad (5.10)
\end{aligned}
$$

because $g(0) = 0$. And for the case $i = 1$, the three points interpolation can not be applied to improve the first block pulse coefficient $\bar{g}_1$ because the functions $f(t)$ and $g(t)$ are not defined for $t < 0$. Therefore the usual two points interpolation of (5.2) is used in this case:

$$\bar{g}_1 = \frac{h}{2}f_1 \qquad (5.11)$$

From the relations (5.9), (5.10) and (5.11), the improvement of the block pulse series of $g(t)$ can be expressed in a vector form:

$$
\begin{aligned}
g(t) &\doteq \left(\begin{array}{cccc} \bar{g}_1 & \bar{g}_2 & \cdots & \bar{g}_m \end{array}\right)\Phi(t) \\
&= \left(\begin{array}{cccc} f_1 & f_2 & \cdots & f_m \end{array}\right)\bar{P}\Phi(t) \qquad (5.12)
\end{aligned}
$$

where the matrix $\bar{P}$ is a modification of the integration operational matrix P:

$$
\bar{P} = \frac{h}{2}\begin{pmatrix}
1 & 2\frac{1}{6} & 2 & \cdots & 2 \\
0 & \frac{5}{6} & 2\frac{1}{6} & \cdots & 2 \\
0 & 0 & \frac{5}{6} & \cdots & 2 \\
\vdots & \vdots & \vdots & \ddots & \vdots \\
0 & 0 & 0 & \cdots & \frac{5}{6}
\end{pmatrix} \qquad (5.13)
$$

Based on this improved integration operational matrix $\bar{P}$, it is easy to express the integral of a function $f(t)$ into its block pulse series:

$$\int_0^t f(t)dt \doteq F^T\bar{P}\Phi(t) \qquad (5.14)$$

In order to show the improvement of the integration operational matrix $\bar{P}$, we compare both the block pulse coefficients $\tilde{g}_i$ and $\bar{g}_i$ ($i = 1, 2, \ldots, m$), which are obtained from $F^T P$ and $F^T \bar{P}$ respectively, with the exact ones g_i, which are directly obtained from $g(t)$ using (1.14). For the case of $\tilde{g}_i$, the difference is:

$$g_i - \tilde{g}_i = \frac{1}{h} \left(\int_{(i-1)h}^{ih} g(t)dt - h \frac{g((i-1)h) + g(ih)}{2} \right) \tag{5.15}$$

From the error analysis of the trapezoidal rule for numerical integration (Atkinson, 1985), we have:

$$\int_{(i-1)h}^{ih} g(t)dt - h \frac{g((i-1)h) + g(ih)}{2} = -\frac{1}{12}h^3 g^{(2)}(\eta) \tag{5.16}$$

for some $\eta \in [(i-1)h, ih]$. Hence

$$g_i - \tilde{g}_i = -\frac{1}{12}h^2 g^{(2)}(\eta) \tag{5.17}$$

For the case of $\bar{g}_i$, since the error analysis of the polynomial interpolation (Atkinson, 1985) gives the difference between $g(t)$ and $\bar{g}(t)$:

$$g(t) - \bar{g}(t) = \frac{(t - (i-2)h)(t - (i-1)h)(t - ih)}{3!} g^{(3)}(\xi) \tag{5.18}$$

for some $\xi \in [(i-2)h, ih]$, after integrating this equation from $(i-1)h$ to ih, we obtain:

$$\int_{(i-1)h}^{ih} g(t)dt - \int_{(i-1)h}^{ih} \bar{g}(t)dt = -\frac{h^4}{24}g^{(3)}(\xi) \tag{5.19}$$

Using (1.14), this equation becomes:

$$g_i - \bar{g}_i = -\frac{h^3}{24}g^{(3)}(\xi) \tag{5.20}$$

Equations (5.17) and (5.20) indicate that the difference $g_i - \bar{g}_i$ may be less than the difference $g_i - \tilde{g}_i$ if the function $g(t)$ is smooth enough and the width of the block pulse h is small. In such cases, the integral operation in the block pulse domain by using the improved operational matrix $\bar{P}$ offers better results than the usual operational matrix P. Here is a numerical example to show this improvement.

Example 5.1 Evaluate the block pulse coefficients of the integral of $f(t) = e^{2t}$ using $F^T P$ and $F^T \bar{P}$ respectively, in the interval $t \in [0,1)$ with $m = 5$.

The evaluation of the requested block pulse coefficients is rather simple. Based on the block pulse coefficients of $f(t)$:

$$F^T = \left(\begin{array}{ccccc} 1.2296 & 1.8343 & 2.7364 & 4.0823 & 6.0901 \end{array} \right)$$

the results can be evaluated from the integration operational matrices P and $\bar{P}$ respectively:

$$\begin{aligned} \tilde{G}^T &= F^T P \\ &= \left(\begin{array}{ccccc} 0.1230 & 0.4293 & 0.8864 & 1.5683 & 2.5855 \end{array} \right) \end{aligned} \tag{5.21}$$

and

$$\bar{G}^T = F^T \bar{P}$$
$$= \left(\begin{array}{ccccc} 0.1230 & 0.4193 & 0.8714 & 1.5459 & 2.5521 \end{array} \right) \tag{5.22}$$

To compare the accuracies of the approximate results in (5.21) and (5.22), we use the same criterion as (4.28), which indicates the sum of the square error between the exact and approximate block pulse coefficients of the results. With the subscripts $j = 1$ for $\tilde{g}(t)$ and $j = 2$ for $\bar{g}(t)$, the criterion gives the values $e_1 = 0.002923$ and $e_2 = 0.000153$, from which the superiority of using the operational matrix $\bar{P}$ is obvious. In this comparison, the exact block pulse coefficients g_k are obtained from the analytical solution:

$$g(t) = \int_0^t f(t)dt = \frac{1}{2}\left(e^{2t} - 1\right)$$

5.2   Generalized integration operational matrices

Multiple integrals appear frequently in the problems of control and systems science. Based on the integration operational matrices P or $\bar{P}$, we can transform these operations into algebraic forms. But the results obtained in such a way have still some deficiencies. Since P^k or $\bar{P}^k$ are powers of integration operational matrices, the computations will increase greatly when the dimension of the matrices becomes large. Since the block pulse function technique is only an approximate approach in the practical uses, errors introduced by the integration operational matrices will be accumulated at each step where the matrices P or $\bar{P}$ are included. These problems influence both the size of computations and the accuracy of results. In order to avoid these deficiencies, one of the ideas is to transform multiple integrals into algebraic operations in one step. This idea can be realized by the generalized integration operational matrices (Wang, 1982).

Consider first the k times integral of a single block pulse function $\phi_i(t)$. According to the Cauchy's integral formula (Oldham, 1974), we have:

$$\underbrace{\int_0^t \cdots \int_0^t}_{k \text{ times}} \phi_i(t)\, dt \cdots dt = \frac{1}{(k-1)!}\int_0^t (t-\tau)^{k-1}\phi_i(\tau)d\tau \tag{5.23}$$

Since the Laplace transform of the above equation is:

$$\mathcal{L}\left\{ \frac{1}{(k-1)!}\int_0^t (t-\tau)^{k-1}\phi_i(\tau)d\tau \right\}$$
$$= \frac{1}{(k-1)!} \cdot \mathcal{L}\left\{t^{k-1}\right\} \cdot \mathcal{L}\left\{\phi_i(t)\right\}$$
$$= \frac{1}{(k-1)!} \cdot \frac{(k-1)!}{s^k} \cdot \frac{1}{s}\left(e^{-(i-1)hs} - e^{-ihs}\right)$$
$$= \frac{1}{s^{k+1}}\left(e^{-(i-1)hs} - e^{-ihs}\right) \tag{5.24}$$

the k times integral of a single block pulse function $\phi_i(t)$ can be written as:

$$\underbrace{\int_0^t \cdots \int_0^t}_{k \text{ times}} \phi_i(t)\, dt \cdots dt$$

$$= \frac{1}{k!}\left((t - (i-1)h)^k \mu(t-(i-1)h) - (t-ih)^k \mu(t-ih)\right) \tag{5.25}$$

where $\mu(t)$ is the unit step function defined in (2.55).

For the first block pulse function $\phi_1(t)$, if the block pulse series of its k times integral is expressed as:

$$\underbrace{\int_0^t \cdots \int_0^t}_{k \text{ times}} \phi_1(t)\, dt \cdots dt \doteq \left(\begin{array}{cccc} c_1 & c_2 & \cdots & c_m \end{array}\right)\Phi(t) \tag{5.26}$$

the block pulse coefficients c_j $(j = 1, 2, \ldots, m)$ can be evaluated from (1.14):

$$\begin{aligned}
c_j &= \frac{1}{h}\int_0^T \frac{1}{k!}\left(t^k \mu(t) - (t-h)^k \mu(t-h)\right)\phi_j(t)dt \\
&= \frac{1}{h}\int_{(j-1)h}^{jh} \frac{1}{k!}\left(t^k \mu(t) - (t-h)^k \mu(t-h)\right)dt \\
&= \begin{cases} \dfrac{h^k}{(k+1)!} & \text{for } j = 1 \\[3mm] \dfrac{h^k}{(k+1)!}\left(j^{k+1} - 2(j-1)^{k+1} + (j-2)^{k+1}\right) & \text{for } j = 2, 3, \ldots, m \end{cases}
\end{aligned} \tag{5.27}$$

For the other block pulse functions $\phi_i(t)$ $(i = 2, 3, \ldots, m)$, the block pulse series of their k times integral can be directly obtained from that of the first block pulse function $\phi_1(t)$ through little modification, because block pulse functions with different subscripts influence only the delay time in (5.25). According to the operation rule of time delay, we have:

$$\underbrace{\int_0^t \cdots \int_0^t}_{k \text{ times}} \phi_i(t)dt \cdots dt \doteq \left(\begin{array}{cccc} c_1 & c_2 & \cdots & c_m \end{array}\right)H^{i-1}\Phi(t)$$

$$= \left(\begin{array}{ccccccc} 0 & \cdots & 0 & \underset{\underset{i\text{th-entry}}{\uparrow}}{c_1} & \cdots & c_{m-i+1} \end{array}\right)\Phi(t) \tag{5.28}$$

From the above discussion, the block pulse series of the k times integrals of all m block pulse functions can be written together in a compact matrix form:

$$\underbrace{\int_0^t \cdots \int_0^t}_{k \text{ times}} \Phi(t)\, dt \cdots dt \doteq P_k \Phi(t) \tag{5.29}$$

where

$$P_k = \frac{h^k}{(k+1)!} \begin{pmatrix} p_{k,1} & p_{k,2} & p_{k,3} & \cdots & p_{k,m} \\ 0 & p_{k,1} & p_{k,2} & \cdots & p_{k,m-1} \\ 0 & 0 & p_{k,1} & \cdots & p_{k,m-2} \\ \vdots & \vdots & \vdots & \ddots & \vdots \\ 0 & 0 & 0 & \cdots & p_{k,1} \end{pmatrix} \qquad (5.30)$$

with

$$p_{k,j} = \begin{cases} 1 & \text{for } j = 1 \\ j^{k+1} - 2(j-1)^{k+1} + (j-2)^{k+1} & \text{for } j = 2, 3, \ldots, m \end{cases} \qquad (5.31)$$

The matrix P_k in (5.30) is defined as the generalized block pulse operational matrix for k times integration, or the kth generalized integration operational matrix. This matrix has also a regular form, i.e. it is an upper triangular matrix, and its kth row can be obtained by shifting the first row $(k-1)$ positions to the right. We can also verify that all its m eigenvalues are $h^k/(k+1)!$. In Table 5.1, we list some values of the parts of entries from (5.31) to have a concrete impression of the generalized integration operational matrices. Obviously, the identity matrix $I = P_0$, the integration operational matrix $P = P_1$ are special cases of the generalized integration operational matrices. To distinguish the integration operational matrix defined in (2.35) from the generalized integration operational matrices, we also call the matrix P as conventional integration operational matrix.

Based on the generalized integration operational matrices, it is easy to express the k times integral of a function $f(t)$ into its block pulse series:

$$\underbrace{\int_0^t \cdots \int_0^t f(t)\,dt \cdots dt}_{k \text{ times}} \doteq F^T P_k \Phi(t) \qquad (5.32)$$

Equation (5.32) shows that the multiple integral operations can be transformed into algebraic operations in one step via the generalized integration operational matrices. Thus, the deficiencies of the conventional integration operational matrix are avoided.

$j=$	1	2	3	4	5	6	7	8
$k=0$	1	0	0	0	0	0	0	0
$k=1$	1	2	2	2	2	2	2	2
$k=2$	1	6	12	18	24	30	36	42
$k=3$	1	14	50	110	194	302	434	590
$k=4$	1	30	180	570	1320	2550	4380	6930
$k=5$	1	62	602	2702	8162	19502	39962	73502

Table 5.1: Entries in generalized integration operational matrices.

Obviously, the computation of using P_k is much smaller than that of using P^k, especially when the numbers m and k are large. The errors generated in the block pulse approximations are also smaller if the generalized integration operational matrices are used. Here is a numerical example to show the reduction of the approximation errors.

Example 5.2 Expand the double integral of $f(t) = t$ into its block pulse series in the interval $t \in [0,1)$ with $m = 4$, based on the conventional and generalized integration operational matrices, respectively.

Since the block pulse coefficient vector of the function $f(t)$ is:

$$F^T = \frac{1}{8} \left(\begin{array}{cccc} 1 & 3 & 5 & 7 \end{array} \right)$$

the conventional integration operational matrix gives the result:

$$\int_0^t \int_0^t f(t) dt dt \doteq F^T P^2 \Phi(t)$$

$$= \frac{1}{1536} \left(\begin{array}{cccc} 3 & 21 & 75 & 189 \end{array} \right) \Phi(t)$$

and the generalized integration operational matrix gives the result:

$$\int_0^t \int_0^t f(t) dt dt \doteq F^T P_2 \Phi(t)$$

$$= \frac{1}{1536} \left(\begin{array}{cccc} 2 & 18 & 70 & 182 \end{array} \right) \Phi(t)$$

But if the double integral of $f(t)$ is expanded directly into its block pulse series according to (1.14), the result is the best one because the minimal mean square error is obtained:

$$\int_0^t \int_0^t f(t) dt dt = \frac{1}{6} t^3$$

$$\doteq \frac{1}{1536} \left(\begin{array}{cccc} 1 & 15 & 65 & 175 \end{array} \right) \Phi(t)$$

Comparing the above three equations, it is clear that the result obtained from the generalized integration operational matrix is better. Therefore the matrix P_k is preferable to the power P^k when the multiple integrals are involved in the problems.

About the block pulse series of successive derivatives based on the generalized integration operational matrices, we can also do the symbolic derivation which is similar to the one based on the conventional integration operational matrix, as discussed in Section 2.2. We can obtain the formula:

$$\frac{d^k g(t)}{dt^k} \doteq \left(G^T P_k^{-1} - \sum_{i=0}^{k-1} g_0^{(i)} E^T P_i P_k^{-1} \right) \Phi(t) \tag{5.33}$$

But in the practical uses of block pulse series expansions, we always try to avoid the matrices P_k^{-1} ($k = 2, 3, \ldots$) in the expressions, because the inverse of the generalized integration operational matrices will lead to more severe divergence problems in the numerical evaluations than the inverse of powers of the conventional integration operational matrices P^{-k}. This is why we always transform differential equations in Chapter 7 first into their equivalent integral equations and then express them into the corresponding algebraic approximations based on the generalized integration operational matrices.

5.3   Properties of generalized integration operational matrices

The generalized integration operational matrices defined in the previous section have some special properties, which can be utilized to simplify calculations when these matrices are involved in the problems. These properties are based on the following relations (Jiang and Schaufelberger, 1985a; Wang and Marleau, 1987):

Relation 5.3.1. For an arbitrary integer l and for the integers $k = 0, 1, \ldots, n - 1$, the following equality holds:

$$\sum_{i=0}^{n}(-1)^{i}\binom{n}{i}(l \pm i)^{k} = 0 \tag{5.34}$$

Relation 5.3.2. For an arbitrary integer l and for the integers $n = 1, 2, \ldots$, the following equality holds:

$$\sum_{i=0}^{n}(-1)^{i}\binom{n}{i}(l + i)^{n} = (-1)^{n}n! \tag{5.35}$$

Relation 5.3.3. For the integers $n = 1, 2, \ldots$, the following equality holds:

$$\sum_{i=0}^{n}\left((-1)^{i}\binom{n}{i}\sum_{s=0}^{n}a_{s}i^{s}\right) = (-1)^{n}a_{n}n! \tag{5.36}$$

where the coefficients a_{s} $(s = 0, 1, \ldots, n)$ are not dependent upon the integer i. In fact, (5.34) and (5.35) are only special cases of (5.36).

Here, we discuss five properties of the generalized integration operational matrices. The first four properties express relations between the entries of the same generalized integration operational matrix P_{k}. But the common factor of the entries $h^{k}/(k + 1)!$ is omitted in the expressions of these properties for simplicity. The fifth property expresses the relation between the different generalized integration operational matrices P_{j} ($j = 1, 2, \ldots, k$).

Property 5.3.1. For the integers $n \geq k \geq 0$ and $l \geq 2$, we have:

$$\sum_{i=0}^{n}(-1)^{i}\binom{n}{i}p_{k,l+n-i} = 0 \tag{5.37}$$

From (5.31), the left-hand side of the above equation becomes:

$$\sum_{i=0}^{n}(-1)^{i}\binom{n}{i}\left[(l + n - i)^{k+1} - 2(l + n - i - 1)^{k+1} + (l + n - i - 2)^{k+1}\right] \tag{5.38}$$

Noticing that the part in the square brackets is a $(k - 1)$th power polynomial with respect to i, this expression equals zero according to (5.36).

In fact, the sum in (5.37) is the nth-order difference of the parts of entries in the same row of the operational matrix P_{k} (Hildebrand, 1974). This difference equals zero so long as the order satisfies $n \geq k$. But the entries of the main diagonal should not be included in this property.

Property 5.3.2. For the integers $n \geq 0$ and $l \geq 2$, we have:

$$\sum_{i=0}^{n}(-1)^i\binom{n}{i}p_{n+1,l+n-i} = (n+2)! \tag{5.39}$$

From (5.31), the left-hand side of the above equation becomes:

$$\sum_{i=0}^{n}(-1)^i\binom{n}{i}\left[(l+n-i)^{n+2} - 2(l+n-i-1)^{n+2} + (l+n-i-2)^{n+2}\right] \tag{5.40}$$

Noticing that the part in the square brackets is an nth power polynomial with respect to i, in which the coefficient of the term i^n is $(-1)^n(n+2)(n+1)$, this expression equals $(n+2)!$ according to (5.36).

Equation (5.39) shows that the nth-order difference of the parts of entries in the same row of the operational matrix P_{n+1} is a constant $(n+2)!$. But the entries of the main diagonal should not be included in this property.

Property 5.3.3. For the integers $n \geq k \geq 0$, we have:

$$\sum_{i=0}^{n}(-1)^i\binom{n}{i}p_{k,n-i+1} = (-1)^{n+k} \tag{5.41}$$

From (5.31), the left-hand side of the above equation becomes:

$$(-1)^n + \sum_{i=0}^{n-1}(-1)^i\binom{n}{i}\left((n-i+1)^{k+1} - 2(n-i)^{k+1} + (n-i-1)^{k+1}\right)$$

$$= (-1)^0\binom{n}{0}(n+1)^{k+1} + (-1)^1\left[2\binom{n}{0} + \binom{n}{1}\right]n^{k+1}$$

$$+ \sum_{i=2}^{n}(-1)^i\left[\binom{n}{i-2} + 2\binom{n}{i-1} + \binom{n}{i}\right](n-i+1)^{k+1}$$

$$= (-1)^{n+k} + \sum_{i=0}^{n+2}(-1)^i\binom{n+2}{i}(n-i+1)^{k+1} \tag{5.42}$$

The second term of this expression equals zero according to (5.34).

Equation (5.41) shows that if the entry of the main diagonal is also involved, the nth-order difference of the parts of entries in the same row of the operational matrix P_k is $(-1)^{n+k}$ so long as the order satisfies $n \geq k$.

Property 5.3.4. For the integers $n-1 \geq k \geq 0$ and $l \geq 1$, we have:

$$\sum_{i=0}^{n}\left((-1)^i\binom{n}{i}\sum_{j=1}^{l+n-i}p_{k,j}\right) = 0 \tag{5.43}$$

Let $b_{n-i} = \sum_{j=1}^{l+n-i} p_{k,j}$ $(i = 0, 1, \ldots, n)$, the left-hand side of (5.43) becomes:

$$\sum_{i=0}^{n} (-1)^i \binom{n}{i} b_{n-i} = \sum_{i=0}^{n-1} (-1)^i \binom{n-1}{i} (b_{n-i} - b_{n-i-1})$$

$$= \sum_{i=0}^{n-1} (-1)^i \binom{n-1}{i} p_{k,l+n-i} \tag{5.44}$$

According to (5.37), the above expression equals zero.

The above equation shows that the first-order differences of the sums b_i $(i = 0, 1, \ldots, n)$ are the parts of entries in the same row of the operational matrix P_k, and their further $(n-1)$th-order differences are zero like Property 5.3.1.

The meaning of the above properties can be explained more clearly by concrete examples. In Figure 5.1, the first lines with values $1, 14, 50, \ldots$ are parts of the entries of the generalized integration operational matrix P_3. The dashed line frames are nth-order differences, with $n = 3, 4$ for Property 5.3.1, with $n = 2$ for Property 5.3.2 and with $n = 3, 4$ for Property 5.3.3, respectively.

Property 5.3.5. If the matrix H_k is defined as:

$$H_k = \frac{h^k}{(k+1)!} \begin{pmatrix} 1 & k & 0 & \cdots & 0 \\ 0 & 1 & k & \cdots & 0 \\ 0 & 0 & 1 & \cdots & 0 \\ \vdots & \vdots & \vdots & \ddots & \vdots \\ 0 & 0 & 0 & \cdots & 1 \end{pmatrix} \tag{5.45}$$

and the matrix G_k $(k = 1, 2, \ldots)$ is defined as:

$$G_k = \frac{h^k}{(k+1)!} \begin{pmatrix} 0 & p_{k,1} & p_{k,2} & \cdots & p_{k,m-1} \\ 0 & 0 & p_{k,1} & \cdots & p_{k,m-2} \\ 0 & 0 & 0 & \cdots & p_{k,m-3} \\ \vdots & \vdots & \vdots & \ddots & \vdots \\ 0 & 0 & 0 & \cdots & 0 \end{pmatrix} \tag{5.46}$$

we have:

$$P_k = H_k + \sum_{j=0}^{k-1} \frac{h^j}{j!} G_{k-j} \tag{5.47}$$

Since for $k = 1, 2, \ldots$, we have:

$$2^{k+1} - 2 \times 1^{k+1} + 0^{k+1} = k + \binom{k+1}{0} + \binom{k+1}{1} + \cdots + \binom{k+1}{k-1} \tag{5.48}$$

and for $k = 1, 2, \ldots$; $i = 2, 3, \ldots, m-1$, we have:

$$(i+1)^{k+1} - 2i^{k+1} + (i-1)^{k+1}$$

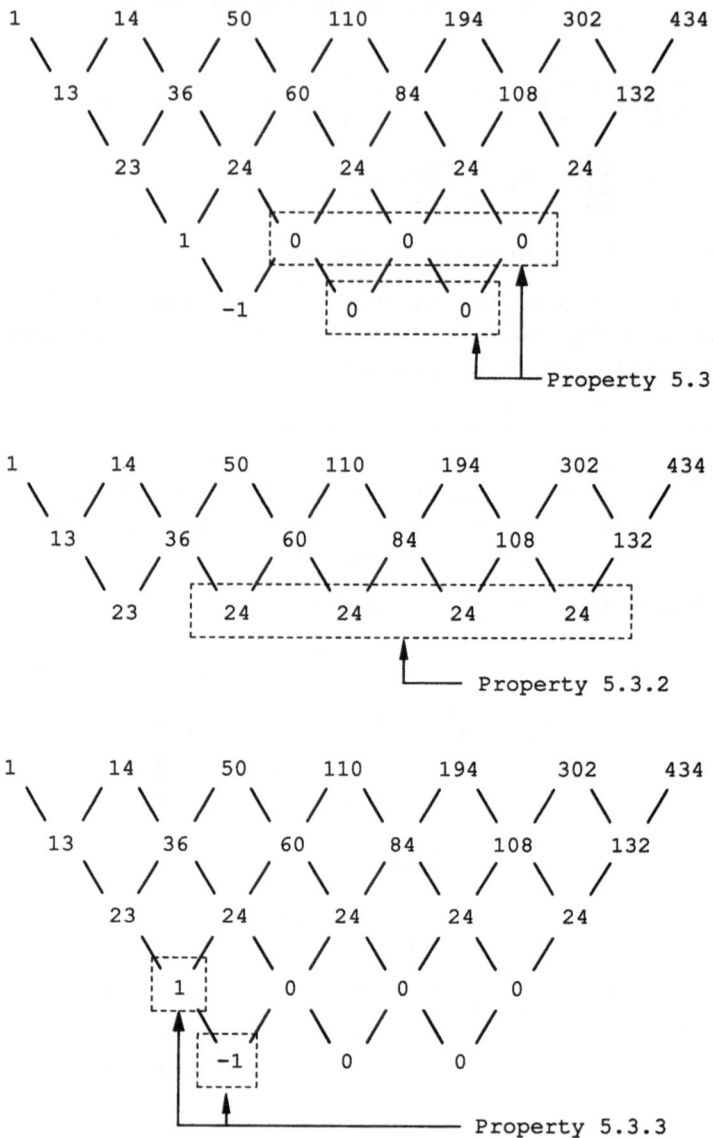

Figure 5.1: Some relations between parts of entries in the matrix P_3.

$$= \sum_{j=0}^{k+1} \binom{k+1}{j} \left(i^{k+1-j} - 2(i-1)^{k+1-j} + (i-2)^{k+1-j} \right)$$

$$= \sum_{j=0}^{k-1} \binom{k+1}{j} \left(i^{k+1-j} - 2(i-1)^{k+1-j} + (i-2)^{k+1-j} \right)$$

$$+\binom{k+1}{k}(i-2(i-1)+(i-2))+\binom{k+1}{k+1}(1-2+1)$$

$$=\sum_{j=0}^{k-1}\binom{k+1}{j}\left(i^{k+1-j}-2(i-1)^{k+1-j}+(i-2)^{k+1-j}\right) \tag{5.49}$$

These two relations can be expressed in a vector form:

$$\begin{pmatrix} 1 \\ 2^{k+1}-2\times 1^{k+1}+0^{k+1} \\ 3^{k+1}-2\times 2^{k+1}+1^{k+1} \\ \vdots \\ m^{k+1}-2(m-1)^{k+1}+(m-2)^{k+1} \end{pmatrix} = \begin{pmatrix} 1 \\ k \\ 0 \\ \vdots \\ 0 \end{pmatrix}$$

$$+\binom{k+1}{0}\begin{pmatrix} 0 \\ 1 \\ 2^{k+1}-2\times 1^{k+1}+0^{k+1} \\ \vdots \\ (m-1)^{k+1}-2(m-2)^{k+1}+(m-3)^{k+1} \end{pmatrix}$$

$$+\binom{k+1}{1}\begin{pmatrix} 0 \\ 1 \\ 2^{k}-2\times 1^{k}+0^{k} \\ \vdots \\ (m-1)^{k}-2(m-2)^{k}+(m-3)^{k} \end{pmatrix}$$

$$+\ \cdots$$

$$+\binom{k+1}{k-1}\begin{pmatrix} 0 \\ 1 \\ 2^{2}-2\times 1^{2}+0^{2} \\ \vdots \\ (m-1)^{2}-2(m-2)^{2}+(m-3)^{2} \end{pmatrix} \tag{5.50}$$

If we neglect the factor $h^k/(k+1)!$ in (5.30), (5.45) and (5.46), the first rows of the matrices P_k, H_k and G_j ($j=1,2,\ldots,k$) are the vectors in (5.50) respectively. Since the other rows of the matrices P_k, H_k and G_j are merely the shifted first rows of the same matrices, we have:

$$P_k = H_k + \binom{k+1}{0}G_k + \frac{h}{k+1}\binom{k+1}{1}G_{k-1}$$

$$+\frac{h^2}{(k+1)k}\binom{k+1}{2}G_{k-2}+\cdots$$

$$+\frac{h^{k-1}}{(k+1)k(k-1)\cdots 3}\binom{k+1}{k-1}G_1$$

$$= H_k + G_k + hG_{k-1} + \frac{h^2}{2!}G_{k-2} + \cdots + \frac{h^{k-1}}{(k-1)!}G_1 \qquad (5.51)$$

The properties discussed above have various applications, e.g. in the recursive evaluation of the entries of the generalized integration operational matrices, in the derivation of the recursive formula for the multiple integrals and in the derivation of the block pulse difference equations with respect to the original differential equations of time-invariant linear systems. Here, we discuss the algorithm for recursively evaluating the entries of the generalized integration operational matrices and the derivation of the recursive formula for multiple integrals. In Chapter 7, we will discuss the problem about the block pulse difference equations from the original differential equations.

At first we consider the problem about the evaluation of entries in the generalized integration operational matrices (Wang and Marleau, 1987). In principle, the entries can be computed directly from the definition formula (5.31). But in practice, it is tedious to do such computations because exponential operations must be applied for each term, especially when the number of the block pulse functions m is chosen large. Therefore, it is necessary to find more efficient algorithms for constructing the generalized integration operational matrices. One such algorithm can be obtained from the properties discussed above.

Since Property 5.3.2 can be expressed in the form:

$$\sum_{i=0}^{k-1}(-1)^i \binom{k-1}{i} p_{k,l-i} = (k+1)! \qquad (5.52)$$

where the integers $k \geq 1$ and $l \geq k + 1$, we obtain directly:

$$p_{k,l} = (k+1)! + \sum_{i=1}^{k-1}(-1)^{i+1}\binom{k-1}{i}p_{k,l-i} \qquad (5.53)$$

Equation (5.53) shows that the term $p_{k,k+1}$ can be evaluated if $p_{k,2}, p_{k,3}, \ldots, p_{k,k}$ are known, and after $p_{k,k+1}$ is obtained, the term $p_{k,k+2}$ can also be evaluated from the $(k-1)$ terms $p_{k,3}, p_{k,4}, \ldots, p_{k,k+1}$. Thus, the recursion can be continued until the last term $p_{k,m}$ is evaluated.

As an example, here is the recursive algorithm for evaluating the entries of the block pulse operational matrix P_3. We evaluate first the terms:

$$p_{3,1} = 1, \qquad p_{3,2} = 14, \qquad p_{3,3} = 50$$

from the definition formula (5.31) as initial values of the recursion. Then we can use the recursive formula (5.53) to evaluate the remaining terms:

$$p_{3,j} = 24 + 2p_{3,j-1} - p_{3,j-2} \qquad (5.54)$$

with $j = 4, 5, \ldots, m$. After all the terms $p_{3,j}$ are obtained, the generalized integration operational matrix P_3 can be constructed together with the common factor $h^3/4!$ according to (5.30).

In comparison with the algorithm of using the definition formula (5.30), the recursive algorithm described above is more efficient in the numerical calculation of the entries of the generalized integration operational matrices. This can be seen from the numbers of multiplications which are applied in the algorithms. If we neglect the multiplications in the factor $h^k/(k+1)!$, the algorithm of using only the definition formula (5.30) contains the following number of multiplications:

$$n_1 = (m-1)(3k+1) \tag{5.55}$$

but the algorithm of using the recursive formula (5.53) contains the following number of multiplications:

$$n_2 = (k-1)(3k+1) + (k-1)(m-k) \tag{5.56}$$

Table 5.2 shows the number of multiplications for various m with $k=3$. It is obvious that the saving of computation is significant in the construction of the generalized integration operational matrices, especially when the number of the block pulse functions m is chosen large for the improvement of the approximate results.

Now we consider the problem about the recursive formula of multiple integrals (Jiang and Schaufelberger, 1985b). In Section 2.4, we derived the recursive formulas for evaluating the block pulse coefficients of integrals based on the conventional integration operational matrix. The formula (2.134) can reduce the size of computations, but it is only suitable for the cases of single integrals. The formula (2.141) can save the size of computations, but it can not avoid the difficulty of accumulative errors for double integrals. In the previous section, we derived the formula (5.32) based on the generalized integration operational matrices. This formula can overcome the difficulty of the accumulative errors for multiple integrals, but the matrix operations should still be done. Thus, the computation size will increase rapidly in such matrix operations, if the number m becomes large. Different from all these cases, the following recursive formula of multiple integrals based on the generalized integration operational matrices can avoid these deficiencies. It can not only overcome the difficulty of accumulative errors, it can also reduce the computation size. From this point of view, this formula is superior to both the recursive formulas based on the conventional integration oper-

	Number of multiplication	m=10	m=20	m=50	m=100
Non-Recursive computation	n_1	90	190	490	990
Recursive computation	n_2	34	54	114	214
	n_2/n_1	37.78 %	28.42 %	23.26 %	21.62 %

Table 5.2: Comparison of number of multiplications in construction of P_3.

ational matrix and the formulas in matrix forms based on the generalized integration operational matrices.

If the function $f(t)$ is k times differentiable in the interval $t \in [0, T]$, we denote it as $f^{(0)}(t)$. We also denote the ith derivative of $f(t)$ as $f^{(i)}(t)$, denote the jth block pulse coefficient and the initial values of the function $f^{(i)}(t)$ as f_j^i and $f_0^{(i)}$, respectively. For the equation:

$$f^{(k)}(t) = \frac{d^k f(t)}{dt^k} \tag{5.57}$$

the k times integration from 0 to t on both sides gives:

$$
\begin{aligned}
f(t) \;=\; & \underbrace{\int_0^t \cdots \int_0^t}_{k \text{ times}} f^{(k)}(t)\, dt \cdots dt \\
& + \underbrace{\int_0^t \cdots \int_0^t}_{(k-1) \text{ times}} f_0^{(k-1)}\, dt \cdots dt \\
& + \quad \cdots \\
& + \int_0^t f_0^{(1)} dt + f_0^{(0)} \tag{5.58}
\end{aligned}
$$

and its block pulse series expansion is:

$$
\begin{aligned}
f(t) \;\doteq\; & \begin{pmatrix} f_1^0 & f_2^0 & \cdots & f_m^0 \end{pmatrix} \Phi(t) \tag{5.59} \\
\doteq\; & \begin{pmatrix} f_1^k & f_2^k & \cdots & f_m^k \end{pmatrix} P_k \Phi(t) \\
& + \begin{pmatrix} f_0^{(k-1)} & f_0^{(k-1)} & \cdots & f_0^{(k-1)} \end{pmatrix} P_{k-1} \Phi(t) \\
& + \quad \cdots \\
& + \begin{pmatrix} f_0^{(1)} & f_0^{(1)} & \cdots & f_0^{(1)} \end{pmatrix} P_1 \Phi(t) \\
& + \begin{pmatrix} f_0^{(0)} & f_0^{(0)} & \cdots & f_0^{(0)} \end{pmatrix} P_0 \Phi(t) \tag{5.60}
\end{aligned}
$$

Similarly, for the equation ($i = 1, 2, \ldots, k-1$):

$$f^{(k)}(t) = \frac{d^{k-i} f^{(i)}(t)}{dt^{k-i}} \tag{5.61}$$

the $(k-i)$ times integration from 0 to t on both sides gives:

$$
\begin{aligned}
f^{(i)}(t) \;=\; & \underbrace{\int_0^t \cdots \int_0^t}_{(k-i) \text{ times}} f^{(k)}(t)\, dt \cdots dt \\
& + \underbrace{\int_0^t \cdots \int_0^t}_{(k-i-1) \text{ times}} f_0^{(k-1)}\, dt \cdots dt \\
& + \quad \cdots \\
& + \int_0^t f_0^{(i+1)} dt + f_0^{(i)} \tag{5.62}
\end{aligned}
$$

and its block pulse series expansions is:

$$
\begin{aligned}
f^{(i)}(t) &\doteq \left(\; f_1^i \quad f_2^i \quad \cdots \quad f_m^i \;\right) \Phi(t) && (5.63)\\
&\doteq \left(\; f_1^k \quad f_2^k \quad \cdots \quad f_m^k \;\right) P_{k-i}\Phi(t)\\
&\quad + \left(\; f_0^{(k-1)} \quad f_0^{(k-1)} \quad \cdots \quad f_0^{(k-1)} \;\right) P_{k-i-1}\Phi(t)\\
&\quad + \cdots\\
&\quad + \left(\; f_0^{(i+1)} \quad f_0^{(i+1)} \quad \cdots \quad f_0^{(i+1)} \;\right) P_1\Phi(t)\\
&\quad + \left(\; f_0^{(i)} \quad f_0^{(i)} \quad \cdots \quad f_0^{(i)} \;\right) P_0\Phi(t) && (5.64)
\end{aligned}
$$

Now consider the equation:

$$
g(t) = f(t) - \sum_{i=0}^{k-1} \frac{h^i}{i!} f^{(i)}(t-h)\mu(t-h) \tag{5.65}
$$

where $\mu(t)$ is the unit step function. Equation (5.65) is the combination of $f^{(i)}(t)$ ($i = 0,1,\ldots,k-1$) with one step time delay h, and it can be expanded into block pulse series according to (5.59), (5.63) and (2.78):

$$
g(t) \doteq \left(\; f_1^0 \quad \left(f_2^0 - \sum_{i=0}^{k-1}\frac{h^i}{i!}f_1^i\right) \quad \cdots \quad \left(f_m^0 - \sum_{i=0}^{k-1}\frac{h^i}{i!}f_{m-1}^i\right) \;\right) \Phi(t) \tag{5.66}
$$

Equation (5.65) can also be expanded into block pulse series in another way according to (5.60), (5.64) and (2.78):

$$
\begin{aligned}
g(t) &\doteq \left(\; f_1^k \quad f_2^k \quad \cdots \quad f_m^k \;\right) \left(P_k - \sum_{i=0}^{k-1}\frac{h^i}{i!}P_{k-i}H\right)\Phi(t)\\
&\quad + \left(\; f_0^{(k-1)} \quad f_0^{(k-1)} \quad \cdots \quad f_0^{(k-1)} \;\right) \left(P_{k-1} - \sum_{i=0}^{k-1}\frac{h^i}{i!}P_{k-i-1}H\right)\Phi(t)\\
&\quad + \cdots\\
&\quad + \left(\; f_0^{(1)} \quad f_0^{(1)} \quad \cdots \quad f_0^{(1)} \;\right) \left(P_1 - \sum_{i=0}^{1}\frac{h^i}{i!}P_{1-i}H\right)\Phi(t)\\
&\quad + \left(\; f_0^{(0)} \quad f_0^{(0)} \quad \cdots \quad f_0^{(0)} \;\right) \left(P_0 - \sum_{i=0}^{0}\frac{h^i}{i!}P_{0-i}H\right)\Phi(t) && (5.67)
\end{aligned}
$$

Using the relation (5.47) and noticing that $G_j = P_j H$ ($j = 1,2,\ldots,k$), the above equation becomes:

$$
\begin{aligned}
g(t) &\doteq \left(\; f_1^k \quad f_2^k \quad \cdots \quad f_m^k \;\right) H_k\Phi(t)\\
&\quad + \left(\; f_0^{(k-1)} \quad f_0^{(k-1)} \quad \cdots \quad f_0^{(k-1)} \;\right) \left(H_{k-1} - \frac{h^{k-1}}{(k-1)!}P_0H\right)\Phi(t)\\
&\quad + \cdots\\
&\quad + \left(\; f_0^{(1)} \quad f_0^{(1)} \quad \cdots \quad f_0^{(1)} \;\right) (H_1 - hP_0H)\Phi(t)\\
&\quad + \left(\; f_0^{(0)} \quad f_0^{(0)} \quad \cdots \quad f_0^{(0)} \;\right) (P_0 - P_0H)\Phi(t)
\end{aligned}
$$

$$= \frac{h^k}{(k+1)!} \left(\begin{array}{cccc} f_1^k & f_2^k + k f_1^k & \cdots & f_m^k + k f_{m-1}^k \end{array} \right) \Phi(t)$$

$$+ \frac{h^{k-1}}{k!} \left(\begin{array}{cccc} f_0^{(k-1)} & 0 & \cdots & 0 \end{array} \right) \Phi(t)$$

$$+ \cdots$$

$$+ \frac{h}{2!} \left(\begin{array}{cccc} f_0^{(1)} & 0 & \cdots & 0 \end{array} \right) \Phi(t)$$

$$+ \left(\begin{array}{cccc} f_0^{(0)} & 0 & \cdots & 0 \end{array} \right) \Phi(t) \qquad (5.68)$$

Comparing (5.68) with (5.66), we can obtain directly the recursive formula for the k times integral of function $f^{(k)}(t)$:

$$f_j^0 = \begin{cases} \dfrac{h^k}{(k+1)!} f_1^k + \displaystyle\sum_{i=0}^{k-1} \dfrac{h^i}{(i+1)!} f_0^{(i)} & \text{for } j = 1 \\[4mm] \dfrac{h^k}{(k+1)!} \left(f_j^k + k f_{j-1}^k \right) + \displaystyle\sum_{i=0}^{k-1} \dfrac{h^i}{i!} f_{j-1}^i & \text{for } j = 2,3,\dots,m \end{cases} \qquad (5.69)$$

Equation (5.69) shows that the block pulse coefficients of function $f^{(0)}(t)$ can be recursively evaluated from the block pulse coefficients of function $f^{(k)}(t)$, if the initial values $f_0^{(0)}$, $f_0^{(1)}$, $\cdots$ and $f_0^{(k-1)}$ are known. In each step for evaluating f_j^0 ($j = 2, 3, \dots, m$), only the data of one previous step f_{j-1}^i ($i = 0, 1, \dots, k$) are used. As a special case with $k = 1$, the recursive formula (5.69) is reduced to (2.134) for single integrals. This is the same as the reduction of the generalized integration operational matrix $P_1 = P$.

According to (5.69), the jth block pulse coefficient of $f(t)$ can be evaluated only when the $(j-1)$th block pulse coefficients of all $f^{(i)}(t)$ ($i = 0, 1, \dots, k-1$) are known. This seems to be a disadvantage of the recursive formula, because a set of intermediate values of block pulse coefficients f_j^i ($i = 1, 2, \dots, k-1; j = 1, 2, \dots, m$) should be evaluated. But in many problems, if the k times integrals are involved, the $k-1, k-2, \dots, 2, 1$ times integrals are also involved. In such cases, this recursive formula can be used more efficiently, because no unnecessary intermediate computations are included. A typical example of such problems is the transform of certain differential equations into their corresponding integral equations, which will be discussed later in Chapter 7.

Although (5.69) indicates that the number of times in the recursion must be restricted to $j \leq m$, we should not take care whether the number j is beyond this restriction or not. The reason is the same as mentioned in Section 2.4, i.e. the number m does not influence the recursive formula, therefore we can assume that the number m is large enough so that the number of times in the recursion always satisfies the restriction of $j \leq m$. This is also an advantage for the practical uses of this recursive formula.

5.4 Discrete integrators and integration operational matrices

In Section 4.1, a close relation between the conventional integration operational matrix and the Tustin integrator (4.5) was obtained. From the Tustin integrator:

$$s^{-1} \doteq \frac{h}{2}\frac{1+z^{-1}}{1-z^{-1}}$$
$$= \frac{h}{2}\left(1 + 2z^{-1} + 2z^{-2} + \cdots\right) \tag{5.70}$$

we can directly construct the conventional integration operational matrix:

$$P = \frac{h}{2}\left(I + 2H + 2H^2 + \cdots\right) \tag{5.71}$$

Here, the terms $H^m, H^{m+1}, \ldots$ are omitted, because the matrix P is an m-dimensional square matrix.

According to (4.5), we can also use the same way to the Madwed integrators to construct the corresponding integration operational matrices. For example, from the third-order Madwed integrator:

$$s^{-3} \doteq \frac{h^3}{24}\frac{1 + 11z^{-1} + 11z^{-2} + z^{-3}}{(1 - z^{-1})^3}$$
$$= \frac{h^3}{24}\left(1 + 14z^{-1} + 50z^{-2} + 110z^{-3} + \cdots\right) \tag{5.72}$$

we obtain a matrix:

$$\frac{h^3}{24}\left(I + 14H + 50H^2 + 110H^3 + \cdots\right) \tag{5.73}$$

Comparing the entries of this matrix with the values related to $k = 3$ in Table 5.1, we notice that the matrix in (5.73) is just the generalized integration operational matrix P_3 which was discussed in Section 5.2. This example shows that the generalized integration operational matrices are related directly to the Madwed integrators. Therefore the reduction of the accumulative errors in multiple integrals via the generalized integration operational matrices can be traced back to the early conclusion of the superiority of the Madwed integrators over the Tustin integrators (Rosko, 1972). Besides, instead of using the recursive formula (5.53), the long division of the Madwed integrators also provides a convenient way to construct the generalized integration operational matrices.

Using the same procedure, we can construct various types of integration operational matrices from other discrete integrators, e.g. from the Boxer-Thaler integrators, the Halijak integrators and so on (Marszalek, 1984c). The different integration operational matrices can be distinguished by the names of the related integrators. For example, the Tustin integration operational matrix means the conventional integration operational matrix, the Madwed integration operational matrices mean the generalized integration operational matrices, and the Boxer-Thaler integration operational matrices mean the integration operational matrices with respect to the Boxer-Thaler integrators. The performance of the integration operational matrices constructed in such a way can also be examined in view of the performance of their corresponding discrete integrators. Since the performance of the discrete integrators has been studied thoroughly in the literature, it is not necessary to discuss the performance of the block pulse operational matrices once more. As an example, Rosko (1972) pointed out: "When the Boxer-Thaler integrating operators are employed to simulate a system, an oscillation or instability may be introduced when the sampling frequency is comparable to the system's natural frequencies". The same phenomenon may also be observed, when the Boxer-Thaler integration operational matrices are used to obtain the block pulse series of the response of a system when the reciprocal of the width of block pulses $1/h$ is comparable to the system's natural frequencies. We show this phenomenon in the following example.

Example 5.3 Find the block pulse series of the unit step response of a fourth-order system with the transfer function:

$$F(s) = \frac{s^2 + 5.5s + 2.5}{s^4 + 5.0s^3 + 9.5s^2 + 8.0s + 2.5}$$

based on the Tustin, Madwed and Boxer-Thaler integration operational matrix, in the interval $t \in [0, 9)$ with the width of block pulses $h = 1$.

For the transfer function of the system:

$$F(s) = \frac{\dfrac{1}{s^2} + \dfrac{5.5}{s^3} + \dfrac{2.5}{s^4}}{1 + \dfrac{5.0}{s} + \dfrac{9.5}{s^2} + \dfrac{8.0}{s^3} + \dfrac{2.5}{s^4}}$$

the block pulse series of the unit step response can be obtained from the Tustin integration operational matrix P:

$$E^T \left(P^2 + 5.5P^3 + 2.5P^4\right) \left(I + 5P + 9.5P^2 + 8P^3 + 2.5P^4\right)^{-1} \Phi(t)$$
$$= \left(\begin{array}{ccccccccc} 0.16 & 0.58 & 0.98 & 1.11 & 1.09 & 1.05 & 1.02 & 1.01 & 1.00 \end{array}\right) \Phi(t)$$

or from the Madwed integration operational matrices P_i $(i = 1, 2, \ldots)$:

$$E^T \left(P_2 + 5.5P_3 + 2.5P_4\right) \left(I + 5P_1 + 9.5P_2 + 8P_3 + 2.5P_4\right)^{-1} \Phi(t)$$
$$= \left(\begin{array}{ccccccccc} 0.08 & 0.69 & 1.01 & 1.08 & 1.07 & 1.04 & 1.02 & 1.01 & 1.00 \end{array}\right) \Phi(t)$$

or from the Boxer-Thaler integration operational matrices G_i $(i = 1, 2, \ldots)$:

$$E^T \left(G_2 + 5.5G_3 + 2.5G_4\right) \left(I + 5G_1 + 9.5G_2 + 8G_3 + 2.5G_4\right)^{-1} \Phi(t)$$
$$= \left(\begin{array}{ccccccccc} 0.02 & 0.91 & 0.61 & 1.70 & 0.02 & 2.81 & -1.93 & 5.96 & -7.30 \end{array}\right) \Phi(t)$$

The results obtained from the Tustin and Madwed integration operational matrices are illustrated in Figure 5.2. Although the width of block pulses is relative large, these results are reasonable. In contrast, the result obtained from the Boxer-Thaler integration

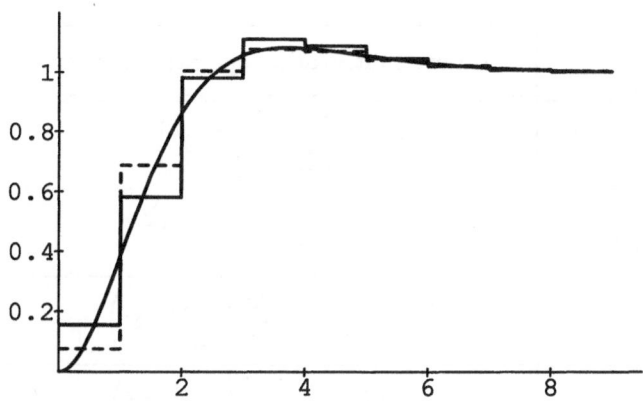

— : from Tustin integrators,   - - - : from Madwed integrators

Figure 5.2: System response approximated by different block pulse series.

operational matrices oscillates divergently. This is just the conclusion mentioned above by Rosko. But with a sufficient small width of block pulses, e.g. $h = 0.2$, all the results obtained from the Tustin, Madwed and Boxer-Thaler integration operational matrices are quite good.

For the construction of the integration operational matrices with respect to discrete integrators of arbitrary orders, no general formulas are found. But some special formulas of them may be useful in practical applications. The construction of the first four Madwed and Boxer-Thaler integration operational matrices are listed in Table 5.3, for the convenience of use.

5.5   Extended integration operational matrices

In this section, we consider the block pulse series of another type of integrals:

$$\underbrace{\int_0^t \cdots \int_0^t t^j f(t)\, dt \cdots dt}_{i\ \text{times}} \tag{5.74}$$

in which the integrands are the products of two functions $f(t)$ and $r(t) = t^j$ with $i, j = 0, 1, \cdots$. The block pulse series of these integrals are useful when block pulse functions are applied to solve problems of time-varying linear systems, which will be discussed later in Chapter 7. In fact, the block pulse series of such integrals are the special cases of Section 2.3 about the integrations of function products. For example, they can be obtained from the combination of the multiplication rule (2.23) and the

	Madwed integration operational matrices
s^{-1}	$h\left(\dfrac{1}{2!}I + \displaystyle\sum_{i=1}^{m-1} H^i\right)$
s^{-2}	$h^2\left(\dfrac{1}{3!}I + \displaystyle\sum_{i=1}^{m-1} iH^i\right)$
s^{-3}	$h^3\left(\dfrac{1}{4!}I + \displaystyle\sum_{i=1}^{m-1} \dfrac{1+6i^2}{12}H^i\right)$
s^{-4}	$h^4\left(\dfrac{1}{5!}I + \displaystyle\sum_{i=1}^{m-1} \dfrac{i+2i^3}{12}H^i\right)$

	Boxer-Thaler integration operational matrices
s^{-1}	$h\left(\dfrac{1}{2}I + \displaystyle\sum_{i=1}^{m-1} H^i\right)$
s^{-2}	$h^2\left(\dfrac{1}{12}I + \displaystyle\sum_{i=1}^{m-1} iH^i\right)$
s^{-3}	$h^3\left(\displaystyle\sum_{i=1}^{m-1} \dfrac{i^2}{2}H^i\right)$
s^{-4}	$h^4\left(-\dfrac{1}{720}I + \displaystyle\sum_{i=1}^{m-1} \dfrac{i^3}{6}H^i\right)$

Table 5.3: First four Madwed and Boxer-Thaler integration operational matrices.

integration rule (2.46):

$$\underbrace{\int_0^t \cdots \int_0^t}_{i \text{ times}} t^j f(t)\,dt \cdots dt \doteq E^T D_F D_R P^i \Phi(t) \tag{5.75}$$

or improved by the generalized integration operational matrices:

$$\underbrace{\int_0^t \cdots \int_0^t}_{i \text{ times}} t^j f(t)\,dt \cdots dt \doteq E^T D_F D_R P_i \Phi(t) \tag{5.76}$$

But in both these equations, the multiplication and integration of functions are approximated by their block pulse series separately. This separation introduces more computations and larger errors in the results. In order to avoid these deficiencies, the integral (5.74) can be expanded into its block pulse series in one step (Hwang and Guo, 1984a).

Consider first the special case of (5.74) with $f(t) = \phi_k(t)$:

$$\underbrace{\int_0^t \cdots \int_0^t}_{i \text{ times}} t^j \phi_k(t)\, dt \cdots dt \qquad (5.77)$$

where $k = 1, 2, \ldots, m$. Since the block pulse functions can be expressed in terms of the delayed unit step functions:

$$\begin{aligned} t^j \phi_k(t) &= t^j \mu(t - (k-1)h) - t^j \mu(t - kh) \\ &= \big((t - (k-1)h) + (k-1)h\big)^j \mu(t - (k-1)h) \\ &\quad - \big((t - kh) + kh\big)^j \mu(t - kh) \end{aligned} \qquad (5.78)$$

the binomial theorem gives:

$$\begin{aligned} t^j \phi_k(t) &= \sum_{q=0}^{j} \binom{j}{q} ((k-1)h)^{j-q} (t - (k-1)h)^q \mu(t - (k-1)h) \\ &\quad - \sum_{q=0}^{j} \binom{j}{q} (kh)^{j-q} (t - kh)^q \mu(t - kh) \end{aligned} \qquad (5.79)$$

and its Laplace transform is:

$$\begin{aligned} \mathcal{L}\{t^j \phi_k(t)\} &= \sum_{q=0}^{j} \binom{j}{q} ((k-1)h)^{j-q} \frac{q!}{s^{q+1}} e^{-(k-1)hs} \\ &\quad - \sum_{q=0}^{j} \binom{j}{q} (kh)^{j-q} \frac{q!}{s^{q+1}} e^{-khs} \end{aligned} \qquad (5.80)$$

According to the integral property of the Laplace transform, we have:

$$\begin{aligned} \mathcal{L}\left\{ \underbrace{\int_0^t \cdots \int_0^t}_{i \text{ times}} t^j \phi_k(t)\, dt \cdots dt \right\} &\\ = \sum_{q=0}^{j} \binom{j}{q} ((k-1)h)^{j-q} \frac{q!}{s^{i+q+1}} e^{-(k-1)hs} &\\ - \sum_{q=0}^{j} \binom{j}{q} (kh)^{j-q} \frac{q!}{s^{i+q+1}} e^{-khs} & \end{aligned} \qquad (5.81)$$

And the inverse Laplace transform of the above equation gives the block pulse series:

$$\underbrace{\int_0^t \cdots \int_0^t}_{i \text{ times}} t^j \phi_k(t)\, dt \cdots dt$$

$$= \sum_{q=0}^{j} \binom{j}{q} ((k-1)h)^{j-q} \frac{q!}{(i+q)!} (t-(k-1)h)^{i+q} \mu(t-(k-1)h)$$

$$- \sum_{q=0}^{j} \binom{j}{q} (kh)^{j-q} \frac{q!}{(i+q)!} (t-kh)^{i+q} \mu(t-kh)$$

$$= \begin{pmatrix} c_{i,j,k,1} & c_{i,j,k,2} & \cdots & c_{i,j,k,m} \end{pmatrix}^{T} \Phi(t) \tag{5.82}$$

where $c_{i,j,k,l}$ ($l = 1, 2, \ldots, m$) are evaluated by (1.14):

$$c_{i,j,k,l} = \frac{1}{h} \underbrace{\int_0^t \cdots \int_0^t}_{(i+1) \text{ times}} t^j \phi_k(t) \, dt \cdots dt \Big|_{t=(l-1)h}^{t=lh}$$

$$= \begin{cases} 0 & \text{for } l < k \\[2mm] h^{i+j} \sum_{q=0}^{j} \binom{j}{q} \frac{q!}{(i+q+1)!} (k-1)^{j-q} & \text{for } l = k \\[4mm] h^{i+j} \sum_{q=0}^{j} \binom{j}{q} \frac{q!}{(i+q+1)!} \Big[(k-1)^{j-q} \big((l-k+1)^{i+q+1} \\[2mm] \quad -(l-k)^{i+q+1} \big) - k^{j-q} \big((l-k)^{i+q+1} - (l-k-1)^{i+q+1} \big) \Big] & \text{for } l > k \end{cases} \tag{5.83}$$

From (5.82), we can express the block pulse series of integrals of all the m block pulse functions together in a matrix form:

$$\underbrace{\int_0^t \cdots \int_0^t}_{i \text{ times}} t^j \Phi(t) \, dt \cdots dt \doteq P_{i,j} \Phi(t) \tag{5.84}$$

where

$$P_{i,j} = \frac{j! h^{i+j}}{(i+j+1)!} \begin{pmatrix} p_{i,j,1,1} & p_{i,j,1,2} & p_{i,j,1,3} & \cdots & p_{i,j,1,m} \\ 0 & p_{i,j,2,2} & p_{i,j,2,3} & \cdots & p_{i,j,2,m} \\ 0 & 0 & p_{i,j,3,3} & \cdots & p_{i,j,3,m} \\ \vdots & \vdots & \vdots & \ddots & \vdots \\ 0 & 0 & 0 & \cdots & p_{i,j,m,m} \end{pmatrix} \tag{5.85}$$

with

$$p_{i,j,k,l} = \begin{cases} 0 & \text{for } l < k \\[2mm] \sum_{q=0}^{j} \frac{(i+j+1)!}{(j-q)!(i+q+1)!} (k-1)^{j-q} & \text{for } l = k \\[4mm] \sum_{q=0}^{j} \frac{(i+j+1)!}{(j-q)!(i+q+1)!} \Big[(k-1)^{j-q} \big((l-k+1)^{i+q+1} \\[2mm] \quad -(l-k)^{i+q+1} \big) - k^{j-q} \big((l-k)^{i+q+1} - (l-k-1)^{i+q+1} \big) \Big] & \text{for } l > k \end{cases} \tag{5.86}$$

The matrix $P_{i,j}$ in (5.85) is defined as the block pulse integration operational matrix related to the i times integral with a term t^j in the integrand, or the ith extended integration operational matrix related to t^j. It is an upper triangular matrix, and its m eigenvalues have the values same as its m entries on the main diagonal. In the above definition, we extract $j!h^{i+j}/(i+j+1)!$ as a common factor in order to keep the remaining parts of the entries $p_{i,j,k,l}$ as integers with $p_{i,j,1,1} = 1$. Obviously, the identity matrix $I = P_{0,0}$, the conventional integration operational matrix $P = P_{1,0}$ and the generalized integration operational matrices $P_i = P_{i,0}$ $(i = 1, 2, \ldots)$ are special cases of this set operational matrices. In Appendix A, we list some examples of the extended integration operational matrices to give a concrete impression.

Based on the operational matrices defined in (5.85) and (5.86), the block pulse series of (5.74) can be obtained in one step:

$$\underbrace{\int_0^t \cdots \int_0^t}_{i \text{ times}} t^j f(t)\, dt \cdots dt \doteq F^T P_{i,j} \Phi(t) \tag{5.87}$$

which can avoid the deficiencies mentioned above. Here is a numerical example to show the improvement of the block pulse series of (5.74) based on the new developed integration operational matrices.

Example 5.4 Find the block pulse series of the integral:

$$g(t) = \int_0^t \int_0^t t^2 e^{-t}\, dt dt$$

in the interval $t \in [0, 5)$ with $m = 10$, based on the formulas (5.75), (5.76) and (5.87), respectively.

In applying the formulas (5.75) and (5.76), both the block pulse coefficients of $f(t) = e^{-t}$ and $r(t) = t^2$ should first be evaluated. After that, we can obtain the block pulse series of the integral from the operational matrices P^2 and P_2 respectively:

$$
\begin{aligned}
g(t) &\doteq E^T D_F D_R P^2 \Phi(t) \\
&= \left(\ 0.0527 \quad 0.3464 \quad 1.0995 \quad 2.3562 \quad 4.0013\ \right) \Phi(t)
\end{aligned}
$$

and

$$
\begin{aligned}
g(t) &\doteq E^T D_F D_R P_2 \Phi(t) \\
&= \left(\ 0.0351 \quad 0.3011 \quad 1.0543 \quad 2.3238 \quad 3.9816\ \right) \Phi(t)
\end{aligned}
$$

But in applying the formula (5.87), it is not necessary to evaluate the block pulse coefficients of the term t^2, the block pulse series expansion is therefore more direct:

$$
\begin{aligned}
g(t) &\doteq F^T P_{2,2} \Phi(t) \\
&= \left(\ 0.0105 \quad 0.2200 \quad 0.9259 \quad 2.1683 \quad 3.8125\ \right) \Phi(t)
\end{aligned}
$$

To compare the accuracies of these approximate results, we use the same criterion as (4.28), which indicates the sum of the square error between the exact and approximate block pulse coefficients of the results. With the subscripts $j = 1$ for (5.75), $j = 2$

for (5.76) and $j = 3$ for (5.87) respectively, the criterion gives the values $e_1 = 0.4714$, $e_2 = 0.3964$ and $e_3 = 0.1356$, from which the superiority of using the operational matrix $P_{2,2}$ is obvious. In this comparison, the exact block pulse coefficients g_k are obtained from the analytical solution:

$$g(t) = -6 + 2t + \left(6 + 4t + t^2\right)e^{-t} \tag{5.88}$$

5.6   Properties of extended integration operational matrices

The extended integration operational matrices defined in the previous section have some special properties, which can be utilized to simplify calculations in the problems. Mainly, these properties express relations between the entries of matrix $P_{i,j}$ in the directions of columns, rows and diagonals, therefore the common factor of the entries $j!h^{i+j}/(i+j+1)!$ can be omitted in the discussions for simplicity. These properties are as follows (Jiang and Schaufelberger, 1991b).

Property 5.6.1. For the integers $i \geq 1$, $j \geq 0$, $d \geq 2$ and $1 \leq r \leq d-1$, we have:

$$\sum_{k=0}^{i+j-1} (-1)^k \binom{i+j-1}{k} p_{i,j,d+i+j-1-k-r,d+i+j-1} = (-1)^{i+1}(i+j+1)! \binom{i+j-1}{j} \tag{5.89}$$

From (5.86), the left-hand side of the above equation becomes:

$$\sum_{k=0}^{i+j-1} \left((-1)^k \binom{i+j-1}{k} \sum_{q=0}^{j} \left(\frac{(i+j+1)!}{(j-q)!(i+q+1)!} \times \right. \right.$$
$$\left[(d+i+j-k-r-2)^{j-q} \left((k+r+1)^{i+q+1} - (k+r)^{i+q+1} \right) \right.$$
$$\left. \left. -(d+i+j-k-r-1)^{j-q} \left((k+r)^{i+q+1} - (k+r-1)^{i+q+1} \right) \right] \right) \right)$$
$$= \sum_{q=0}^{j} \left(\frac{(i+j+1)!}{(j-q)!(i+q+1)!} \sum_{k=0}^{i+j-1} \left((-1)^k \binom{i+j-1}{k} \times \right. \right.$$
$$\left[(d+i+j-k-r-2)^{j-q} \left((k+r+1)^{i+q+1} - (k+r)^{i+q+1} \right) \right.$$
$$\left. \left. -(d+i+j-k-r-1)^{j-q} \left((k+r)^{i+q+1} - (k+r-1)^{i+q+1} \right) \right] \right) \right) \tag{5.90}$$

Noticing that the part in the square brackets is an $(i+j-1)$th power polynomial with respect to k in which the coefficient of the term k^{i+j-1} is $(-1)^{j-q}(i+q+1)(i+j)$, the above expression can be simplified according to (5.36):

$$(-1)^{i+1}(i+j+1)! \sum_{q=0}^{j} (-1)^q \frac{(i+j)!}{(j-q)!(i+q)!}$$

$$= (-1)^{i+j+1}(i+j+1)! \sum_{q=0}^{j}(-1)^q \binom{i+j}{q}$$

$$= (-1)^{i+1}(i+j+1)! \binom{i+j-1}{j} \tag{5.91}$$

Equation (5.89) shows that the $(i+j-1)$th-order difference of the entries in the same column of the operational matrix $P_{i,j}$ is a constant. This constant is dependent only upon the integers i and j, no matter which column is used. But the entries of the main diagonal should not be included in the $(i+j-1)$th-order difference. This property is illustrated by the example of the operational matrix $P_{3,2}$ in Figure 5.3, where the fourth-order differences are all 4320.

Property 5.6.2. For the integers $i \geq 0$, $j \geq 0$, $n \geq i+j$, $d \geq 2$ and $1 \leq r \leq d-1$, we have:

$$\sum_{k=0}^{n}(-1)^k \binom{n}{k} P_{i,j,d+n-k-r,d+n} = 0 \tag{5.92}$$

For the case of $i = 0$, the equality (5.92) is obvious, because the operational matrices $P_{0,j}$ are diagonal. For the case of $i > 0$, the constant value of the $(i+j-1)$th-order difference in Property 5.6.1 gives automatically a zero value of the differences with order higher than $(i+j-1)$. This property is also illustrated by the example of the operational matrix $P_{3,2}$ in Figure 5.3, where the fifth-order differences are all zero.

Property 5.6.3. For the integers $i \geq 0$, $j \geq 0$ and $d \geq 1$, we have:

$$\sum_{k=0}^{j}(-1)^k \binom{j}{k} P_{i,j,d+j-k,d+j-k} = \frac{(i+j+1)!}{(i+1)!} \tag{5.93}$$

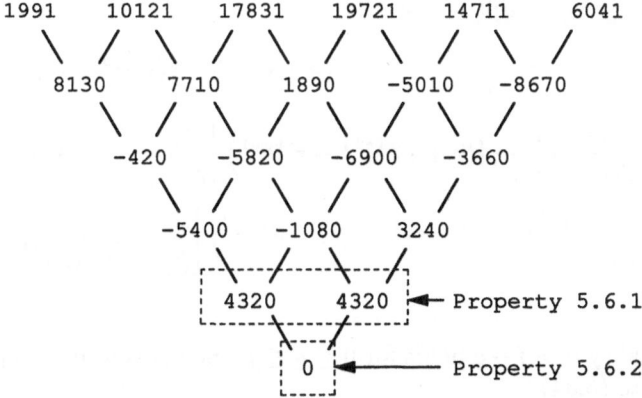

Figure 5.3: Differences of parts of entries in the same column of matrix $P_{3,2}$.

From (5.86), the left-hand side of the above equation becomes:

$$
\sum_{k=0}^{j}\left((-1)^k\binom{j}{k}\sum_{q=0}^{j}\left(\frac{(i+j+1)!}{(j-q)!(i+q+1)!}(d+j-k-1)^{j-q}\right)\right)
$$

$$
= \frac{(i+j+1)!}{j!(i+1)!}\left[\sum_{k=0}^{j}(-1)^k\binom{j}{k}(d+j-k-1)^j\right]
$$

$$
+ \sum_{q=1}^{j}\left(\frac{(i+j+1)!}{(j-q)!(i+q+1)!}\left[\sum_{k=0}^{j}(-1)^k\binom{j}{k}(d+j-k-1)^{j-q}\right]\right) \qquad (5.94)
$$

This expression equals $(i+j+1)!/(i+1)!$ because the sum in the first square brackets is $j!$ according to (5.35) and the sums in the second square brackets are zero according to (5.34).

Equation (5.93) shows that the jth-order difference of the entries of the main diagonal of the operational matrix $P_{i,j}$ is a constant. This constant is dependent only upon the integers i and j, no matter which part of the main diagonal is used in calculating the difference. This property is illustrated by the example of the operational matrix $P_{3,2}$ in Figure 5.4, where the second-order differences are all 30.

Property 5.6.4. For the integers $i \geq 0$, $j \geq 0$, $n \geq j+1$, $d \geq 1$, and $r \geq 0$, we have:

$$
\sum_{k=0}^{n}(-1)^k\binom{n}{k}p_{i,j,d+n-k,d+n-k+r} = 0 \qquad (5.95)
$$

For the subscript $r = 0$, we have the case of the main diagonal. Since the jth-order difference of the entries on the main diagonal line of the operational matrix $P_{i,j}$ is constant, the differences with order higher than j are automatically zero. For the subscript $r > 0$, this is the case for the remaining diagonals. From (5.86), the left-hand side of the above equation becomes:

$$
\sum_{k=0}^{n}\left((-1)^k\binom{n}{k}\sum_{q=0}^{j}\left(\frac{(i+j+1)!}{(j-q)!(i+q+1)!}\left[(d+n-k-1)^{j-q}\left((r+1)^{i+q+1}-r^{i+q+1}\right)\right.\right.\right.
$$

$$
\left.\left.\left.-(d+n-k)^{j-q}\left(r^{i+q+1}-(r-1)^{i+q+1}\right)\right]\right)\right)
$$

$$
= \sum_{q=0}^{j}\left(\frac{(i+j+1)!}{(j-q)!(i+q+1)!}\left((r+1)^{i+q+1}-r^{i+q+1}\right)\left[\sum_{k=0}^{n}(-1)^k\binom{n}{k}(d+n-k-1)^{j-q}\right]\right)
$$

$$
-\sum_{q=0}^{j}\left(\frac{(i+j+1)!}{(j-q)!(i+q+1)!}\left(r^{i+q+1}-(r-1)^{i+q+1}\right)\left[\sum_{k=0}^{n}(-1)^k\binom{n}{k}(d+n-k)^{j-q}\right]\right)
$$

$$
(5.96)
$$

Since the inequality $n > j - q$ holds for $0 \leq q \leq j$, the sums in both square brackets are zero according to (5.34).

Equation (5.95) shows that the nth-order differences of the entries on the same diagonal line of the operational matrix $P_{i,j}$ are zero, so long as the order satisfies $n > j$.

This property is illustrated by the example of the operational matrix $P_{3,2}$ in Figure 5.4, where the third-order differences are all zero.

Property 5.6.5. For the integers $i \geq 0$, $j \geq 0$, $n \geq i$, $d \geq 1$ and $r \geq 1$, we have:

$$\sum_{k=0}^{n}(-1)^k \binom{n}{k}p_{i,j,d,d+n-k+r} = 0 \tag{5.97}$$

For the case of $i = 0$, the equality (5.97) is obvious, because the operational matrices $P_{0,j}$ are diagonal. For the case of $i > 0$, we first study the integral:

$$f_{i,k}(t) = \underbrace{\int_0^t \cdots \int_0^t f(t)\phi_k(t)\,dt \cdots dt}_{i \text{ times}} \tag{5.98}$$

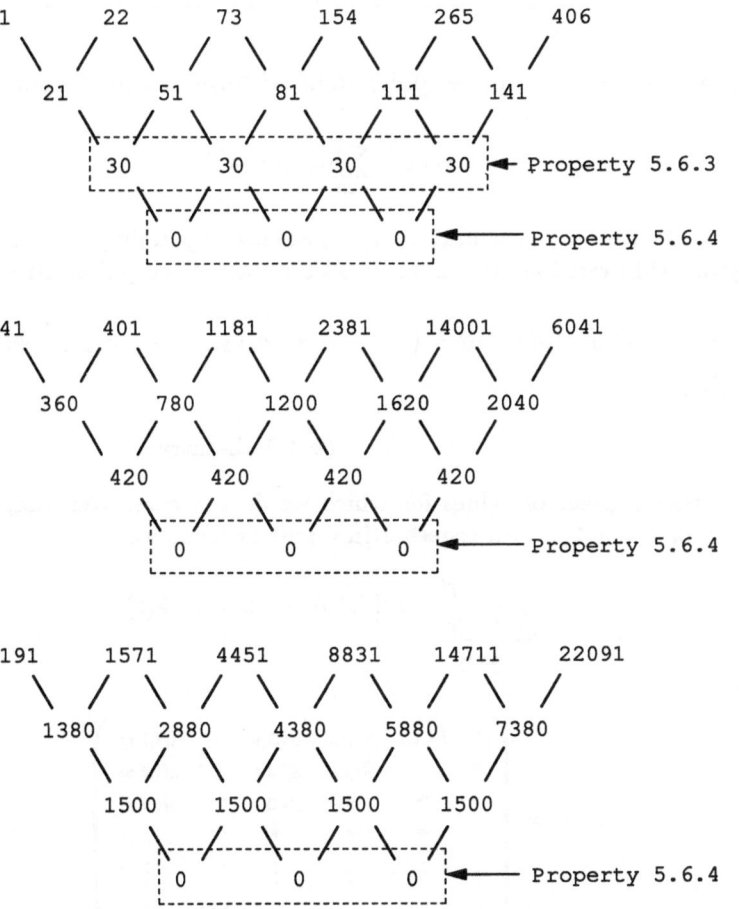

Figure 5.4: Differences of parts of entries in the same diagonal of matrix $P_{3,2}$.

where $f(t)$ is an arbitrary continuous function and $\phi_k(t)$ is the kth block pulse function. Since the integrand $f(t)\phi_k(t)$ takes a non-zero value only in the interval $t \in [(k-1)h, kh)$, the solution of the integral (5.98) can be found in the interval $t \in [kh, T)$, which is an $(i-1)$th power polynomial with respect to $(t-kh)$:

$$g_{i,k}(t) = \begin{cases} f_{i,k}(t) & \text{for } 0 \le t < kh \\ \sum_{s=1}^{i} \dfrac{x_{i-s+1,k}}{(s-1)!}(t-kh)^{s-1} & \text{for } kh \le t < T \end{cases} \tag{5.99}$$

where the constants are $x_{j,k} = f_{j,k}(kh) - f_{j,k}(0)$ $(j = 1, 2, \ldots, i)$. According to (1.14), the lth block pulse coefficient for $l > k$ can be evaluated as:

$$\begin{aligned} g_{i,k,l} &= \frac{1}{h} \int_{(l-1)h}^{lh} g_{i,k}(t) dt \\ &= \sum_{s=1}^{i} \frac{h^{s-1} x_{i-s+1,k}}{s!} \left((l-k)^s - (l-k-1)^s \right) \end{aligned} \tag{5.100}$$

Obviously, it is an $(i-1)$th power polynomial with respect to $(l-k)$:

$$g_{i,k,l} = \sum_{s=0}^{i-1} a_{k,s}(l-k)^s \tag{5.101}$$

where the coefficients $a_{k,s}$ $(s = 0, 1, \ldots, i-1)$ are not depending upon the integer $(l-k)$. The integral (5.98) can then be approximated by the block pulse series:

$$\underbrace{\int_0^t \cdots \int_0^t f(t)\phi_k(t)\, dt \cdots dt}_{i \text{ times}} \doteq \left(\ast \quad \cdots \quad \ast \quad g_{i,k,k+1} \quad \cdots \quad g_{i,k,m} \right) \Phi(t) \tag{5.102}$$

$$\underset{(k+1)\text{th-entry}}{\uparrow}$$

where $\ast$ is used in place of values for which we do not care. Rearranging the m integrals (5.98) for $k = 1, 2, \ldots, m$ together in a matrix form, we get:

$$\underbrace{\int_0^t \cdots \int_0^t f(t)\Phi(t)\, dt \cdots dt}_{i \text{ times}} \doteq G_i \Phi(t) \tag{5.103}$$

where

$$G_i = \begin{pmatrix} \ast & g_{i,1,2} & g_{i,1,3} & g_{i,1,4} & \cdots & g_{i,1,m} \\ \ast & \ast & g_{i,2,3} & g_{i,2,4} & \cdots & g_{i,2,m} \\ \ast & \ast & \ast & g_{i,3,4} & \cdots & g_{i,3,m} \\ \ast & \ast & \ast & \ast & \cdots & g_{i,4,m} \\ \vdots & \vdots & \vdots & \vdots & \ddots & \vdots \\ \ast & \ast & \ast & \ast & \cdots & \ast \end{pmatrix} \tag{5.104}$$

In fact, the matrix G_i has a zero lower triangular part, because the integrand $f(t)\phi_k(t)$ is zero in the interval $t \in [0, (k-1)h)$.

We observe the following relation between the entries of the same row in the matrix G_i with integers $n \geq i$, $d \geq 1$ and $r \geq 1$:

$$\sum_{k=0}^{n}(-1)^k \binom{n}{k} g_{i,d,d+n-k+r} = \sum_{k=0}^{n}\left((-1)^k \binom{n}{k}\left[\sum_{s=0}^{i-1} a_{d,s}(n-k+r)^s\right]\right) \qquad (5.105)$$

Noticing that the part in the square brackets is an $(i-1)$th power polynomial with respect to k, the above expression equals zero according to (5.36). This general observation yields directly the relation between the entries of the same row in the operational matrices $P_{i,j}$ ($i = 1, 2, \ldots; j = 0, 1, \ldots$) because the function t^j and the matrix $P_{i,j}$ in (5.84) are only special cases of the function $f(t)$ and the matrix G_i in (5.103).

Equation (5.97) shows that the nth-order difference of the entries in the same row of the operational matrix $P_{i,j}$ equals zero so long as the order satisfies $n \geq i$. But the entries of the main diagonal should not be included in the difference. This property is illustrated by the example of the operational matrix $P_{3,2}$ in Figure 5.5, where the third-order differences are all zero.

Property 5.6.6. For the integers $i \geq 1$, $j \geq 0$, $d \geq 1$ and $r \geq 0$, we have:

$$\sum_{l=0}^{i-1}\left((-1)^l \binom{i-1}{l}\sum_{k=0}^{j}\left((-1)^k \binom{j}{k} P_{i,j,d+j-k,d+j-k+i-l+r}\right)\right) = (i+j+1)! \qquad (5.106)$$

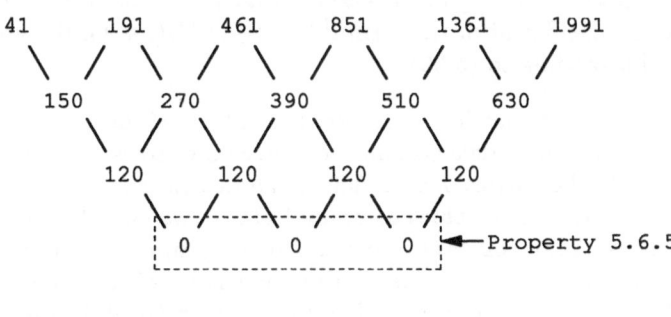

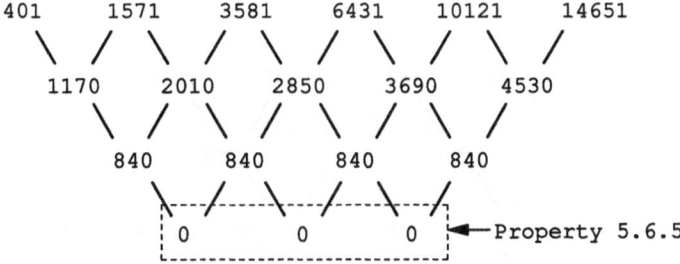

Figure 5.5: Differences of parts of entries in the same row of matrix $P_{3,2}$.

From (5.86), the left-hand side of the above equation becomes:

$$\sum_{l=0}^{i-1}\left((-1)^l\binom{i-1}{l}\sum_{k=0}^{j}\left((-1)^k\binom{j}{k}\sum_{q=0}^{j}\frac{(i+j+1)!}{(j-q)!(i+q+1)!}\times\right.\right.$$

$$\left((d+j-k-1)^{j-q}\left((i-l+r+1)^{i+q+1}-(i-l+r)^{i+q+1}\right)\right.$$

$$\left.\left.\left.-(d+j-k)^{j-q}\left((i-l+r)^{i+q+1}-(i-l+r-1)^{i+q+1}\right)\right)\right)\right)$$

$$=\sum_{l=0}^{i-1}\sum_{q=0}^{j}\left((-1)^l\binom{i-1}{l}\frac{(i+j+1)!}{(j-q)!(i+q+1)!}\times\right.$$

$$\left(\left[\sum_{k=0}^{j}(-1)^k\binom{j}{k}(d+j-k-1)^{j-q}\right]\left((i-l+r+1)^{i+q+1}-(i-l+r)^{i+q+1}\right)\right.$$

$$\left.\left.-\left[\sum_{k=0}^{j}(-1)^k\binom{j}{k}(d+j-k)^{j-q}\right]\left((i-l+r)^{i+q+1}-(i-l+r-1)^{i+q+1}\right)\right)\right)$$

$$(5.107)$$

According to (5.34) and (5.35), the sums in both square brackets are $j!$ for $q=0$ and are zero for $q=1,2,\ldots,j$, the above expression can be simplified to:

$$\frac{(i+j+1)!}{(i+1)!}\sum_{l=0}^{i-1}(-1)^l\binom{i-1}{l}\left[(i-l+r+1)^{i+1}-2(i-l+r)^{i+1}+(i-l+r-1)^{i+1}\right] \quad (5.108)$$

Noticing that the part in the square brackets is an $(i-1)$th power polynomial with respect to l, and the coefficient of the term l^{i-1} is $(-1)^{i+1}i(i+1)$, the above expression equals $(i+j+1)!$ according to (5.36).

Equation (5.106) shows that if we consider the entries of the operational matrix $P_{i,j}$ on the $(j+1)$ successive diagonals and on the i successive rows, after first applying the jth-order difference to the entries of each diagonal and then applying the $(i-1)$th-order difference to the i intermediate values computed above, we can obtain a constant from these two differences. This constant is dependent only upon the integers i and j, no matter which diagonals and rows are used. But the entries of the main diagonal should not be included in the two differences. The property in (5.106) is illustrated by the

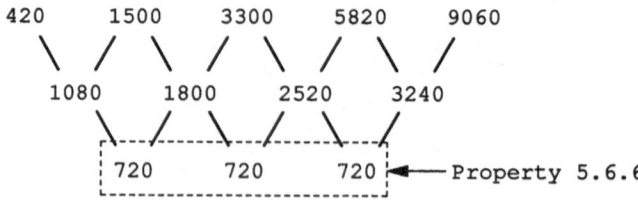

Figure 5.6: Differences of intermediate results in (5.106).

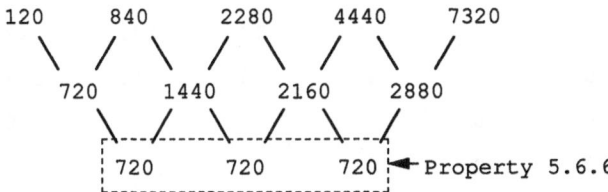

Figure 5.7: Differences of intermediate results in (5.109).

example of the operational matrix $P_{3,2}$ in Figures 5.4 and 5.6. Figure 5.4 gives the second-order differences, i.e. the jth-order differences, in the direction of the diagonals, and Figure 5.6 gives the second-order differences, i.e. the $(i-1)$th-order differences, of the values obtained above. The constants are all 720.

Equation (5.106) can also be rewritten as:

$$\sum_{k=0}^{j}\left((-1)^k\binom{j}{k}\sum_{l=0}^{i-1}\left((-1)^l\binom{i-1}{l}P_{i,j,d+j-k,d+j-k+i-l+r}\right)\right)=(i+j+1)! \qquad (5.109)$$

This means that if we consider the entries of the operational matrix $P_{i,j}$ on the i successive rows and on the $(j+1)$ successive diagonals, after first applying the $(i-1)$th-order difference to the entries of each row and then applying the jth-order difference to the $(j+1)$ intermediate values computed above, we can also obtain the same constant from these two differences.

The property in (5.109) is illustrated by the same example of $P_{3,2}$ in Figure 5.5 and 5.7. Figure 5.5 gives the second-order differences, i.e. the $(i-1)$th-order differences, in the direction of the rows, and Figure 5.7 gives the second-order differences, i.e. the jth-order differences, of the values obtained above. The constants are still 720.

Property 5.6.7. For the integers $i \geq 0$, $j \geq 0$, $n \geq i+j+1$ and $d \geq 1$, we have:

$$\sum_{k=0}^{n}\left((-1)^k\binom{n}{k}\sum_{s=1}^{d+n-k}P_{i,j,s,d+n-k}\right)=0 \qquad (5.110)$$

For the subscript $s = d+n-k$, we have the case of the main diagonal. This part equals zero according to (5.95). Thus from (5.86), the left-hand side of the above equation becomes:

$$\sum_{k=0}^{n}\left((-1)^k\binom{n}{k}\sum_{s=1}^{d+n-k-1}\sum_{q=0}^{j}\frac{(i+j+1)!}{(j-q)!(i+q+1)!}\times\right.$$
$$\left[(s-1)^{j-q}\left((d+n-k-s+1)^{i+q+1}-(d+n-k-s)^{i+q+1}\right)\right.$$
$$\left.\left.-s^{j-q}\left((d+n-k-s)^{i+q+1}-(d+n-k-s-1)^{i+q+1}\right)\right]\right) \qquad (5.111)$$

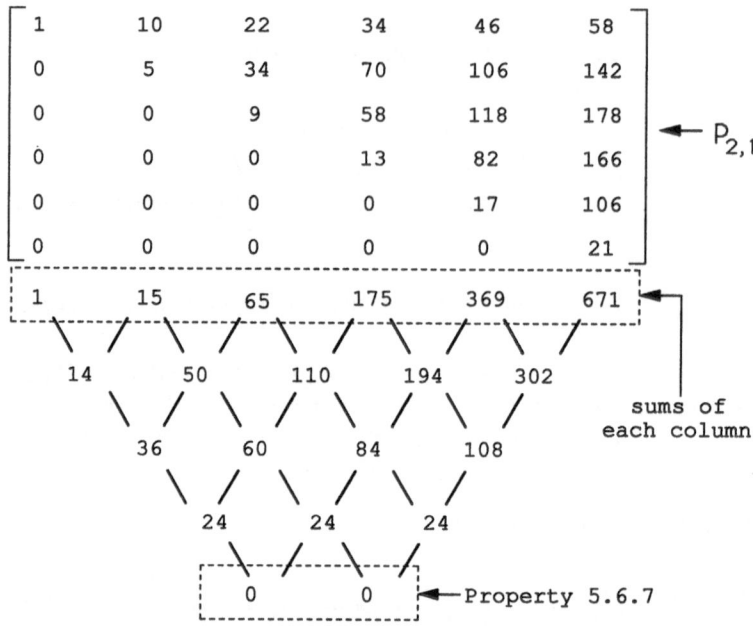

Figure 5.8: Differences of entries of successive columns in matrix $P_{2,1}$.

Noticing that the part in the square brackets is an $(i + j)$th power polynomial with respect to k, the left-hand side of (5.111) equals zero according to (5.36) for $n > i + j$.

Since the lower triangular parts of the block pulse operational matrices are zeros, (5.110) shows that if we consider entries of the operational matrix $P_{i,j}$ on the $(n + 1)$ successive columns, after first adding all the entries in each column separately and then applying the nth-order difference to the $(n + 1)$ intermediate sums computed above, the result is zero so long as the order satisfies $n > i + j$. This property is illustrated by the example of the operational matrix $P_{2,1}$ in Figure 5.8, where the fourth-order differences of the intermediate sums are applied.

As special cases of the extended operational matrices, the generalized integration operational matrices are $P_{i,0}$, which are upper triangular matrices and their second to mth rows are merely the shifted first rows to the right. In these cases, the properties discussed above are also simplified. With $j = 0$, Properties 5.6.3, 5.6.4 become trivial, Properties 5.6.2, 5.6.5 become the same as (5.37), Properties 5.6.1, 5.6.6 become the same as (5.39), and Property 5.6.7 also becomes the same as (5.43).

The properties discussed above have various applications, e.g. in the recursive evaluation of the entries of the extended integration operational matrices and in the derivation of the block pulse regression equations with respect to the original differential equations of time-varying linear systems. Here, we discuss the algorithm for recursively evaluating the entries of the extended integration operational matrices. In Chapter 7, we will discuss the problem about the block pulse regression equations from the original differential equations.

In principle, the entries of the extended integration operational matrices can be computed directly from the definition (5.86). But in practice, it is tedious to do such computations because exponential operations must be applied for each term, especially when the number of the block pulse functions m is chosen large. Therefore, it is necessary to find more efficient algorithms for the construction of the extended integration operational matrices. Since the properties discussed above reveal the relations between the entries in the diagonals, rows and columns, they can be combined flexibly with each other for the recursive computation of the entries in these integration operational matrices. Here is one of such algorithms for the construction of the extended integration operational matrices which is based on the combination of Properties 5.6.4 and 5.6.5.

As the first step with respect to the rows, the entries $p_{i,j,d,s}$ ($s = d+i+1, d+i+2, \ldots, m$) can be evaluated recursively from (5.97):

$$p_{i,j,d,s} = \sum_{k=1}^{i}(-1)^{k+1}\binom{i}{k}p_{i,j,d,s-k} \tag{5.112}$$

Equation (5.112) shows that the entry $p_{i,j,d,s}$ can be computed if the values of the previous entries $p_{i,j,d,s-i}, p_{i,j,d,s-i+1}, \ldots, p_{i,j,d,s-1}$ are known. This implies that the first i entries of each row should be evaluated directly from the definition formula (5.86). But only $i \times (j+1)$ entries must be computed in this way because we evaluate the entries of the first $(j+1)$ rows in this step. As the second step with respect to each diagonal, the entries $p_{i,j,d,s}$ ($d = j+2, j+3, \ldots, m; s = d+1, d+2, \ldots, m$) can be evaluated recursively from (5.95):

$$p_{i,j,d,s} = \sum_{k=1}^{j+1}(-1)^{k+1}\binom{j+1}{k}p_{i,j,d-k,s-k} \tag{5.113}$$

Equation (5.113) shows that the entry $p_{i,j,d,s}$ can be computed if the values of the previous entries $p_{i,j,d-j-1,s-j-1}, p_{i,j,d-j,s-j}, \ldots, p_{i,j,d-1,s-1}$ are known. Since all the entries of the first $(j + 1)$ rows have already been evaluated in the first step, the recursive computation of (5.113) can be done immediately. Using these two steps, the operational matrix $P_{i,j}$ can be constructed numerically.

As an example, here is the recursive formula for evaluating the entries of the block pulse operational matrix $P_{3,2}$. Figure 5.9 illustrates the above computational algorithm. In the first step, we evaluate first the parts of the entries:

$p_{3,2,1,1} = 1,$	$p_{3,2,1,2} = 41,$	$p_{3,2,1,3} = 191,$	$p_{3,2,1,4} = 461,$
$p_{3,2,2,2} = 22,$	$p_{3,2,2,3} = 401,$	$p_{3,2,2,4} = 1571,$	$p_{3,2,2,5} = 3581,$
$p_{3,2,3,3} = 73,$	$p_{3,2,3,4} = 1181,$	$p_{3,2,3,5} = 4451,$	$p_{3,2,3,6} = 10001$

from the definition formula (5.86) as initial values of the recursion. We can then use the recursive formula (5.112) to evaluate the parts of the remaining entries of the first three rows:

$$p_{3,2,d,s} = 3p_{3,2,d,s-1} - 3p_{3,2,d,s-2} + p_{3,2,d,s-3} \tag{5.114}$$

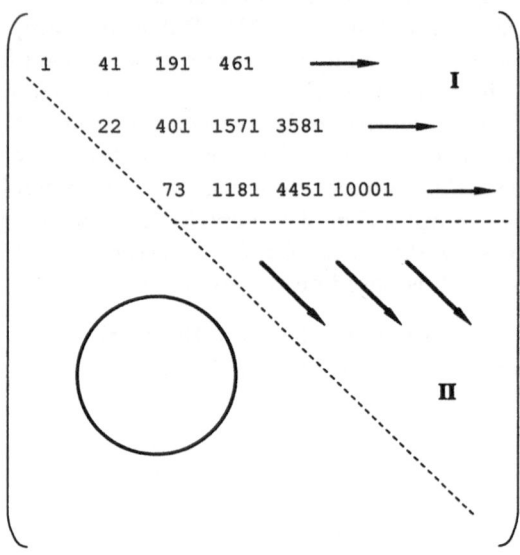

Figure 5.9: Computational algorithm with two steps (5.112) and (5.113).

with $d = 1, 2, 3$ and $s = d + 4, d + 5, \ldots, m$. In the second step, we can directly use the recursive formula (5.113) to evaluate the remaining unknown entries:

$$p_{3,2,d,s} = 3p_{3,2,d-1,s-1} - 3p_{3,2,d-2,s-2} + p_{3,2,d-3,s-3} \qquad (5.115)$$

with $d = 4, 5, \ldots, m$ and $s = d + 1, d + 2, \ldots, m$. After inserting the common factor $h^5/360$, the operational matrix $P_{3,2}$ is numerically constructed.

In comparison with the algorithm of using the definition formula (5.86), the recursive algorithm described above is more efficient in the numerical calculation of the entries of block pulse operational matrices. This can be seen from the numbers of multiplications which are involved in the algorithms. If we neglect the multiplications in the

	Number of multiplication	m=10	m=20	m=50	m=100
Non-recursive computation	n_1	2412	10127	65072	262647
Recursive computation	n_2	612	1077	4272	15597
	n_2/n_1	25.37 %	10.63 %	6.57 %	5.94 %

Table 5.4: Comparison of number of multiplications in construction of $P_{3,2}$.

factor $j!h^{i+j}/(j-q)!/(i+q+1)!$, the non-recursive algorithm contains the number of multiplications:

$$
n_1 = \underbrace{(m-1) \times \frac{j(j+1)}{2}}_{\text{main diagonal part}}
$$

$$
+ \underbrace{\frac{m(m-1)}{2} \times (3j^2 + 4ij + 3i + 3j + 2)}_{\text{upper triangular part}} \tag{5.116}
$$

In contrast, the recursive algorithm with two steps contains the number of multiplications:

$$
n_2 = \underbrace{j \times \frac{j(j+1)}{2} + i(j+1) \times (3j^2 + 4ij + 3i + 3j + 2) + \frac{(j+1)(2m - 2i - j - 2)}{2} \times i}_{\text{in the first step}}
$$

$$
+ \underbrace{\frac{(m-j-1)(m-j)}{2} \times (j+1)}_{\text{in the second step}} \tag{5.117}
$$

Table 5.4 gives the number of multiplications for various m in the numerical calculation of the operational matrix $P_{3,2}$. It is obvious that the reduction of computational effort in the recursive algorithm is significant, especially when the number m becomes larger.

Chapter 6

Nonparametric representations of dynamic systems

In the time domain, impulse responses are usually applied to describe dynamic characteristics of continuous-time linear systems. Using convolution integrals, system outputs can be determined by the system inputs based on the known system impulse responses. Such a way of representation for dynamic systems can also be used in the block pulse domain. In this chapter, after the block pulse transfer matrices of both time-invariant and time-varying linear systems are introduced, simple relations between system inputs and outputs can be found, and block diagrams of linear systems can be established.

6.1 Block pulse transfer matrices of linear systems

For a time-invariant linear system, if its impulse response is described by the function $g(t)$, its input $u(t)$ and output $y(t)$ under zero initial values can be associated by the convolution integral:

$$y(t) = \int_0^t u(\tau)g(t - \tau)d\tau \tag{6.1}$$

This convolution integral is usually applied as the nonparametric representation of a time-invariant linear system. In this case, the impulse response is a curve which contains information about the dynamic characteristics of the system.

In the block pulse domain, similar nonparametric representations of time-invariant linear systems can also be found. As discussed in Section 2.2 about the operation rule of convolution integral, if U, Y and G are the block pulse coefficient vectors of the input, output and impulse response of a system respectively, (6.1) leads directly to a relation in the block pulse domain:

$$Y^T \doteq U^T J_G \tag{6.2}$$

where the matrix J_G is constructed by the vector G according to (2.62). This relation associates the system input and output in a much simpler way, because no convolution

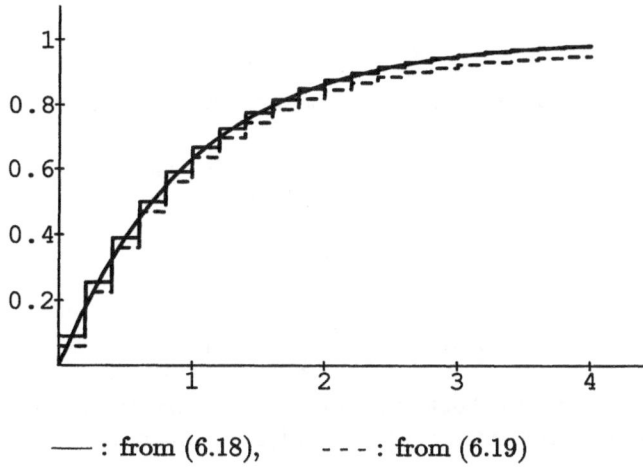

$$— : \text{from } (6.18), \qquad - - - : \text{from } (6.19)$$

Figure 6.2: Output approximations of a time-invariant linear system.

Together with the analytical solution of the output $y(t) = 1 - e^{-t}$, the two variations of the piecewise constant approximate results are illustrated in Figure 6.2. It is clear that the result obtained from (2.124) is worse than the one obtained from (2.61). This extra error is introduced by the main diagonal entries of the matrix in (2.125), which take the value $g_1 h/3$ instead of $g_1 h/2$. This difference will influence each block pulse coefficient of the result, so long as the block pulse coefficients of the system input do not take the value of zero. This example indicates that the operation rule (2.61) is preferable to (2.124) when the convolution integrals are involved in the problems.

Example 6.2 Find the block pulse transfer matrix of a time-varying linear system in the interval $t \in [0, 8)$ with $m = 20$, and evaluate the block pulse coefficients of the output $y(t)$ from the obtained block pulse transfer matrix if the input is $u(t) = e^{-t/2}$. The system is constructed by a RC lowpass filter and an amplitude modulator, as shown in Figure 6.3 where the symbol $\otimes$ stands for a multiplicator. The filter is the same as the one in Example 6.1, and the modulating signal is $x(t) = \cos t$.

If $\delta(t)$ is a unit impulse function and excites the modulator at the time instant τ, the response of the modulator is $\delta(t - \tau) \cos \tau$. Therefore the impulse response of the

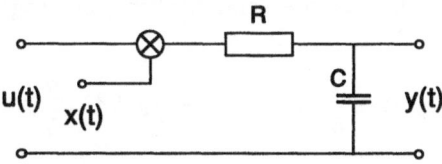

Figure 6.3: A time-varying linear system in Example 6.2.

modulator together with the RC lowpass filter is:

$$g(t, \tau) = e^{-(t-\tau)} \cos \tau \tag{6.20}$$

From (6.8), the block pulse coefficients of $g(t, \tau)$ can be evaluated as:

$$g_k(t) = \frac{1}{h} \int_{(k-1)h}^{kh} g(t, \tau) d\tau$$

$$= \frac{e^{-t}}{h} f_k$$

where

$$f_k = \frac{e^\tau (\cos \tau + \sin \tau)}{2} \bigg|_{(k-1)h}^{kh}$$

and the block pulse transfer matrix of this time-varying linear system can be obtained from (6.13), in which the entries $g_{k,j}^{(t)}$ $(k, j = 1, 2, \ldots, m)$ are:

$$g_{k,j}^{(t)} = \begin{cases} \dfrac{e^{-(j-1)h}}{h} \left(1 - e^{-h}\right) f_k & \text{for } j > k \\[2ex] \dfrac{e^{-(k-1)h}}{h^2} \left(1 - e^{-h}(1+h)\right) f_k & \text{for } j = k \\[2ex] 0 & \text{for } j < k \end{cases} \tag{6.21}$$

Concretely, the block pulse transfer matrix of $t \in [0, 8)$ with $m = 20$ is:

$$G^{(t)} = \begin{pmatrix} 0.1837 & 0.2638 & 0.1768 & 0.1185 & \cdots \\ 0 & 0.1537 & 0.2207 & 0.1480 & \cdots \\ 0 & 0 & 0.0994 & 0.1428 & \cdots \\ 0 & 0 & 0 & 0.0295 & \cdots \\ \vdots & \vdots & \vdots & \vdots & \ddots \end{pmatrix}$$

After the block pulse coefficients of the input are evaluated:

$$U^T = \begin{pmatrix} 0.9063 & 0.7421 & 0.6075 & 0.4974 & \cdots \end{pmatrix}$$

we have the piecewise constant approximate result of the output:

$$Y^T \doteq U^T G^{(t)}$$

$$= \begin{pmatrix} 0.1665 & 0.3532 & 0.3845 & 0.3186 & \cdots \end{pmatrix} \tag{6.22}$$

It is illustrated in Figure 6.4 as the solid broken line, together with the analytical solution of the output:

$$y(t) = \frac{2}{5} (\cos t + 2 \sin t) e^{-t/2} - \frac{2}{5} e^{-t}$$

In fact, this example is somewhat special, because we can obtain the analytical expressions of both the impulse response of the system (6.20) and the entries of the block

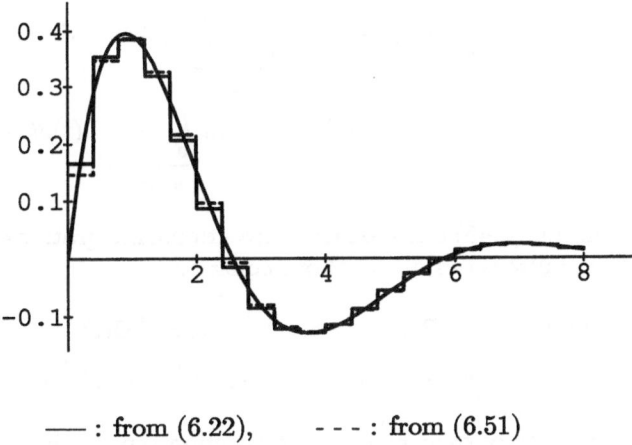

— : from (6.22),   - - - : from (6.51)

Figure 6.4: Output approximations of a time-varying linear system.

pulse transfer matrix (6.21). But for time-varying linear systems, it is generally not so easy to obtain both these analytical expressions. Therefore the establishment of the block pulse transfer matrices from the system impulse responses is practically difficult. In order to overcome this difficulty, we can use other methods to construct the block pulse transfer matrices of systems, e.g. using the relation between block pulse transfer matrices and differential equations, or obtaining such nonparametric representations through identification. These methods will be discussed in the next sections.

6.2 Block pulse transfer matrices and differential equations

For both time-invariant and time-varying linear systems, the block pulse transfer matrices can also be obtained rather easily from the known differential equations, although differential equations are usually used as parametric models of systems.

Consider a time-invariant linear system which is described by a differential equation:

$$y^{(n)}(t) + a_{n-1}y^{(n-1)}(t) + \cdots + a_1 y^{(1)}(t) + a_0 y(t)$$
$$= b_n u^{(n)}(t) + \cdots + b_1 u^{(1)}(t) + b_0 u(t) \tag{6.23}$$

where $u^{(k)}(t)$ and $y^{(k)}(t)$ are the kth derivatives of the input $u(t)$ and output $y(t)$, respectively. Although certain terms may not exist in this differential equation (e.g. the direct coupling between the input and output signals usually does not exist in physical systems), we still use this general form for the convenience of expressions. Integrating this differential equation n times from 0 to t on both sides under zero initial values, we have:

$$y(t) + a_{n-1} \int_0^t y(t) + \cdots + a_0 \underbrace{\int_0^t \cdots \int_0^t}_{n \text{ times}} y(t) dt \cdots dt$$

$$= b_n u(t) + b_{n-1} \int_0^t u(t) + \cdots + b_0 \underbrace{\int_0^t \cdots \int_0^t}_{n \text{ times}} u(t) dt \cdots dt \qquad (6.24)$$

Using the operation rule (2.46) based on the conventional integration operational matrix, we obtain the block pulse series of the above equation:

$$Y^T \left(I + a_{n-1} P + \cdots + a_1 P^{n-1} + a_0 P^n \right) \Phi(t)$$

$$\doteq U^T \left(b_n I + b_{n-1} P + \cdots + b_1 P^{n-1} + b_0 P^n \right) \Phi(t) \qquad (6.25)$$

Equating the coefficients of each block pulse function separately on both sides of the above equation, we have:

$$Y^T A = U^T B \qquad (6.26)$$

where

$$A = I + a_{n-1} P + \cdots + a_1 P^{n-1} + a_0 P^n \qquad (6.27)$$

and

$$B = b_n I + b_{n-1} P + \cdots + b_1 P^{n-1} + b_0 P^n \qquad (6.28)$$

If A is nonsingular, the block pulse coefficient vector of the output can be determined by that of the input:

$$Y^T = U^T B A^{-1} \qquad (6.29)$$

Comparing (6.29) with (6.3), the block pulse transfer matrix of a time-invariant linear system can be constructed by:

$$G^{(t)} = B A^{-1} \qquad (6.30)$$

Consider a time-varying linear system with the differential equation:

$$y^{(n)}(t) + a_{n-1}(t) y^{(n-1)}(t) + \cdots + a_1(t) y^{(1)}(t) + a_0(t) y(t)$$

$$= b_n(t) u^{(n)}(t) + \cdots + b_1(t) u^{(1)}(t) + b_0(t) u(t) \qquad (6.31)$$

Using the same procedure as described above, after integrating this differential equation n times from 0 to t on both sides under zero initial values, we have:

$$y(t) + \int_0^t a_{n-1}(t) y(t) + \cdots + \underbrace{\int_0^t \cdots \int_0^t}_{n \text{ times}} a_0(t) y(t) dt \cdots dt$$

$$= b_n(t) u(t) + \int_0^t b_{n-1}(t) u(t) + \cdots + \underbrace{\int_0^t \cdots \int_0^t}_{n \text{ times}} b_0(t) u(t) dt \cdots dt \qquad (6.32)$$

According to the operation rules (2.26) and (2.46), each term in this equation can be expanded into its block pulse series, for example:

$$\underbrace{\int_0^t \cdots \int_0^t}_{k \text{ times}} a_{n-k}(t)y(t)dt \cdots dt$$

$$\doteq \underbrace{\int_0^t \cdots \int_0^t}_{k \text{ times}} Y^T \Phi(t)\Phi^T(t)A_{n-k}dt \cdots dt$$

$$= \underbrace{\int_0^t \cdots \int_0^t}_{k \text{ times}} Y^T D_{A_{n-k}}\Phi(t)dt \cdots dt$$

$$\doteq Y^T D_{A_{n-k}} P^k \Phi(t) \tag{6.33}$$

where A_{n-k} is the block pulse coefficient vector of the function $a_{n-k}(t)$. Therefore the block pulse series of the equation is:

$$Y^T \left(I + D_{A_{n-1}}P + \cdots + D_{A_1}P^{n-1} + D_{A_0}P^n\right)\Phi(t)$$
$$\doteq U^T \left(D_{B_n} + D_{B_{n-1}}P + \cdots + D_{B_1}P^{n-1} + D_{B_0}P^n\right)\Phi(t) \tag{6.34}$$

Equating the coefficients of each block pulse function separately on both sides of the above equation, we have:

$$Y^T A = U^T B \tag{6.35}$$

where

$$A = I + D_{A_{n-1}}P + \cdots + D_{A_1}P^{n-1} + D_{A_0}P^n \tag{6.36}$$

and

$$B = D_{B_n} + D_{B_{n-1}}P + \cdots + D_{B_1}P^{n-1} + D_{B_0}P^n \tag{6.37}$$

If A is nonsingular, the block pulse coefficient vector of the output can be determined by that of the input:

$$Y^T = U^T B A^{-1} \tag{6.38}$$

Comparing (6.38) with (6.12), the block pulse transfer matrix of a time-varying linear system can also be constructed by:

$$G^{(t)} = B A^{-1} \tag{6.39}$$

In fact, the matrices A, B in (6.27), (6.28) for time-invariant linear systems are only special forms of the matrices A, B in (6.36) and (6.37) for time-varying linear systems, because in the time-invariant case, the diagonal matrices are reduced to identity matrices multiplied by constants, e.g. D_{A_k} is reduced to $a_k I$. Therefore the block pulse transfer matrix $G^{(t)}$ constructed by (6.36), (6.37) and (6.39) can be applied to both time-invariant and time-varying linear systems.

Since the multiple integrals can also be expanded into their block pulse series based on the generalized integration operational matrices, for example:

$$\underbrace{\int_0^t \cdots \int_0^t}_{k \text{ times}} a_{n-k}(t)y(t)dt \cdots dt$$

$$\doteq \underbrace{\int_0^t \cdots \int_0^t}_{k \text{ times}} Y^T D_{A_{n-k}} \Phi(t) dt \cdots dt$$

$$\doteq Y^T D_{A_{n-k}} P_k \Phi(t) \tag{6.40}$$

the block pulse transfer matrix $G^{(t)}$ can be constructed in this way to improve the accuracy of the approximations. The only change in the above discussion is to substitute the generalized integration operatoral matrices P_k for the corresponding powers of the conventional integration operatoral matrix P^k in the matrices A and B, i.e. (6.36) and (6.37) are now improved to:

$$A = I + D_{A_{n-1}} P_1 + \cdots + D_{A_1} P_{n-1} + D_{A_0} P_n \tag{6.41}$$

and

$$B = D_{B_n} + D_{B_{n-1}} P_1 + \cdots + D_{B_1} P_{n-1} + D_{B_0} P_n \tag{6.42}$$

In constructing block pulse transfer matrices from the corresponding system differential equations, a lot of matrix operations are involved, e.g. matrix multiplications appear in the computations of A, B in (6.36), (6.37), and matrix inversions appear in the computations of $G^{(t)}$ in (6.39). If the time interval under study is large, especially if the input and output signals oscillate a lot, we must unavoidably choose a rather large number m to obtain meaningful approximations of results which take piecewise constant values. But this large number m leads to large computations in the matrix operations. Fortunately, all the matrices involved in the establishment of block pulse transfer matrices have upper triangular forms, and this feature can be utilized to save computations. Using the formulas in Appendix B, we can construct the block pulse transfer matrices without matrix operations. Noticing that all the matrices A, B and $G^{(t)}$ are special upper triangular matrices in the case of time-invariant linear systems, (6.39) can be simplified by the computation of the entries:

$$g_i^{(t)} = \begin{cases} \dfrac{1}{a_1} b_i & \text{for } i = 1 \\[3mm] \dfrac{1}{a_1} \left(b_i - \displaystyle\sum_{k=1}^{i-1} g_k^{(t)} a_{i-k+1} \right) & \text{for } i = 2, 3, \ldots, m \end{cases} \tag{6.43}$$

to construct the matrix:

$$G^{(t)} = \begin{pmatrix} g_1^{(t)} & g_2^{(t)} & \cdots & g_m^{(t)} \\ 0 & g_1^{(t)} & \cdots & g_{m-1}^{(t)} \\ \vdots & \vdots & \ddots & \vdots \\ 0 & 0 & \cdots & g_1^{(t)} \end{pmatrix} \tag{6.44}$$

And noticing that all the matrices A, B and $G^{(t)}$ are general upper triangular matrices in the case of time-varying linear systems, (6.39) can be simplified by the computation of the entries:

$$g_{i,j}^{(t)} = \begin{cases} 0 & \text{for } j < i \\ \dfrac{1}{a_{j,j}} b_{i,j} & \text{for } j = i \\ \dfrac{1}{a_{j,j}} \left(b_{i,j} - \sum_{k=i}^{j-1} g_{i,k}^{(t)} a_{k,j} \right) & \text{for } j > i \end{cases} \tag{6.45}$$

to construct the matrix:

$$G^{(t)} = \begin{pmatrix} g_{1,1}^{(t)} & g_{1,2}^{(t)} & \cdots & g_{1,m}^{(t)} \\ 0 & g_{2,2}^{(t)} & \cdots & g_{2,m}^{(t)} \\ \vdots & \vdots & \ddots & \vdots \\ 0 & 0 & \cdots & g_{m,m}^{(t)} \end{pmatrix} \tag{6.46}$$

Using the formulas of this section, we solve the problems of Examples 6.1 and 6.2 once more. In Example 6.1, since the RC lowpass filter is described by a differential equation:

$$y^{(1)}(t) + y(t) = u(t) \tag{6.47}$$

its block pulse transfer matrix $G^{(t)}$ can be evaluated from (6.39):

$$\begin{aligned} G^{(t)} &= P(I+P)^{-1} \\ &= \begin{pmatrix} 0.0909 & 0.1653 & 0.1352 & 0.1106 & \cdots \\ 0 & 0.0909 & 0.1653 & 0.1352 & \cdots \\ 0 & 0 & 0.0909 & 0.1653 & \cdots \\ 0 & 0 & 0 & 0.0909 & \cdots \\ \vdots & \vdots & \vdots & \vdots & \ddots \end{pmatrix} \end{aligned} \tag{6.48}$$

and the piecewise constant approximate result of the output is:

$$\begin{aligned} Y^T &\doteq U^T G^{(t)} \\ &= \begin{pmatrix} 0.0909 & 0.2562 & 0.3914 & 0.5021 & \cdots \end{pmatrix} \end{aligned} \tag{6.49}$$

The differences between the results obtained here and those in (6.16), (6.18) are small. In Example 6.2, since the filter together with an amplitude modulator is described by a differential equation:

$$y^{(1)}(t) + y(t) = x(t)u(t) \tag{6.50}$$

and the block pulse coefficients of $x(t)$ is:

$$X^T = \begin{pmatrix} 0.9735 & 0.8198 & 0.5367 & 0.1688 & \cdots \end{pmatrix}$$

the block pulse transfer matrix $G^{(t)}$ can also be evaluated from (6.39):

$$G^{(t)} = D_X P(I+P)^{-1}$$

$$= \begin{pmatrix} 0.1623 & 0.2704 & 0.1803 & 0.1202 & \cdots \\ 0 & 0.1366 & 0.2277 & 0.1518 & \cdots \\ 0 & 0 & 0.0895 & 0.1491 & \cdots \\ 0 & 0 & 0 & 0.0281 & \cdots \\ \vdots & \vdots & \vdots & \vdots & \ddots \end{pmatrix}$$

the piecewise constant approximate result of the output is:

$$Y^T \doteq U^T G^{(t)}$$
$$= \begin{pmatrix} 0.1471 & 0.3465 & 0.3867 & 0.3262 & \cdots \end{pmatrix} \qquad (6.51)$$

To compare with the early result in (6.22), this result is also shown in Figure 6.4 as the dotted broken line.

6.3 Block pulse transfer matrices and system block diagrams

In control theory and systems science, block diagrams are applied extensively to describe structures of systems, because relations between system inputs and outputs can be seen clearly from them. For time-invariant linear systems, their dynamics can be expressed by the corresponding transfer functions in the block diagrams. But if some parts of systems are time-varying, their dynamics should be expressed in other ways, e.g. by their impulse responses.

In the previous sections, nonparametric representations of linear systems were discussed in the block pulse domain. Similar to system transfer functions and system impulse responses, block pulse transfer matrices can also be applied to express dynamics of systems in the block diagrams. But in the way of using block pulse transfer matrices, we need not distinguish the time-invariant and time-varying cases, and we can avoid the convolution integral operations. This may be seen as an advantage of the nonparametric representations in the block pulse domain, as shown in Figure 6.5.

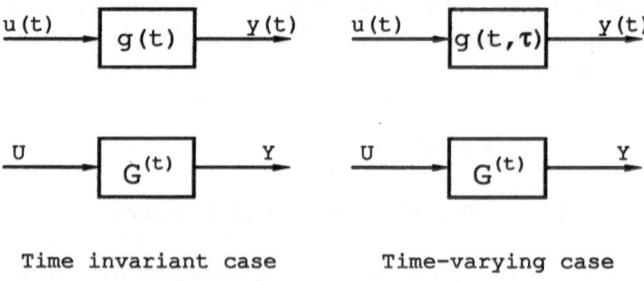

Figure 6.5: Block diagram of linear systems.

In the following, we will discuss the block pulse transfer matrices of some elementary linear systems. These discussions outline the principle of using block pulse transfer matrices in the block diagrams, and the elementary linear systems can also be applied further in constructing block pulse transfer matrices of more complicated systems. In the figures below, the block diagrams of the corresponding time-invariant linear systems are also given for the purpose of comparison with the expressions using transfer functions.

1. Cascade systems. The block diagram of a cascade system is illustrated in Figure 6.6. From the relations:

$$X^T = U^T G_1^{(t)} \tag{6.52}$$

and

$$Y^T = X^T G_2^{(t)} \tag{6.53}$$

we have:

$$Y^T = U^T G_1^{(t)} G_2^{(t)} \tag{6.54}$$

Equation (6.54) shows that the block pulse transfer matrix of a cascade system is the product of block pulse transfer matrices of subsystems:

$$G^{(t)} = G_1^{(t)} G_2^{(t)} \tag{6.55}$$

Generally speaking, block pulse transfer matrices of subsystems which are involved in a cascade system are not interchangeable in the multiplications. But if all the subsystems are time-invariant, their block pulse transfer matrices have special upper triangular forms and can be interchanged with each other according to Appendix B.

2. Parallel systems. The block diagram of a parallel system is illustrated in Figure 6.7. From the relations:

$$X_1^T = U^T G_1^{(t)} \tag{6.56}$$

$$X_2^T = U^T G_2^{(t)} \tag{6.57}$$

and

$$Y^T = X_1^T + X_2^T \tag{6.58}$$

we have:

$$Y^T = U^T \left(G_1^{(t)} + G_2^{(t)} \right) \tag{6.59}$$

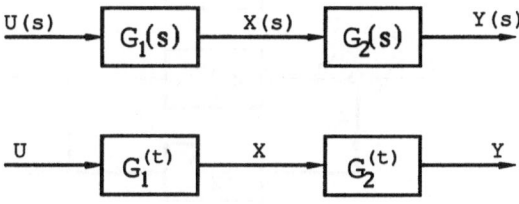

Figure 6.6: Block diagram of a cascade system.

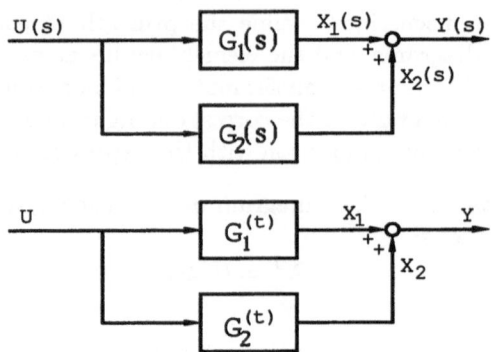

Figure 6.7: Block diagram of a parallel system.

Equation (6.59) shows that the block pulse transfer matrix of a parallel system is the sum of block pulse transfer matrices of subsystems:

$$G^{(t)} = G_1^{(t)} + G_2^{(t)} \tag{6.60}$$

3. Simple feedback systems. The block diagram of a simple feedback system is illustrated in Figure 6.8. From the relations:

$$X_1^T = U^T - X_2^T \tag{6.61}$$

$$X_2^T = Y^T G_2^{(t)} \tag{6.62}$$

and

$$Y^T = X_1^T G_1^{(t)} \tag{6.63}$$

we have:

$$X_1^T \left(I + G_1^{(t)} G_2^{(t)} \right) = U^T \tag{6.64}$$

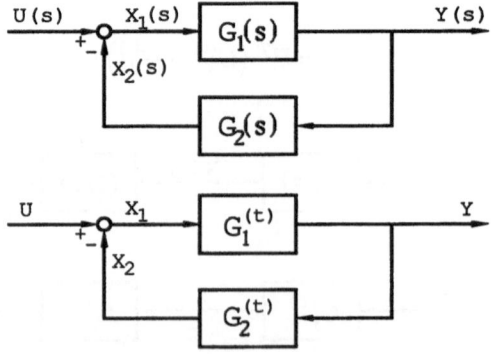

Figure 6.8: Block diagram of a feedback system.

If the matrix $(I + G_1^{(t)} G_2^{(t)})$ is not singular, we obtain:

$$Y^T = U^T \left(I + G_1^{(t)} G_2^{(t)}\right)^{-1} G_1^{(t)} \tag{6.65}$$

Equation (6.65) shows that the block pulse transfer matrix of a simple feedback system is:

$$G^{(t)} = \left(I + G_1^{(t)} G_2^{(t)}\right)^{-1} G_1^{(t)} \tag{6.66}$$

which can also be written as:

$$G^{(t)} = G_1^{(t)} \left(I + G_2^{(t)} G_1^{(t)}\right)^{-1} \tag{6.67}$$

through the identical transformation of matrices.

4. Systems containing time delay. There are two cases of time delay in the systems, i.e. the delay in the input side and the delay in the output side. Both these cases are illustrated in the block diagrams of Figure 6.9, where the time delay is $\tau = (q + \lambda)h$ with a nonnegative integer q and a fractional part $0 \leq \lambda < 1$.

For the case of time delay in the input side, from the relations:

$$Y^T = X^T G_1^{(t)} \tag{6.68}$$

and

$$X^T = U^T \left((1 - \lambda)H^q + \lambda H^{q+1}\right) \tag{6.69}$$

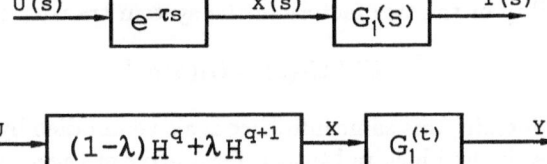

Case 1. Input signal contains time delay

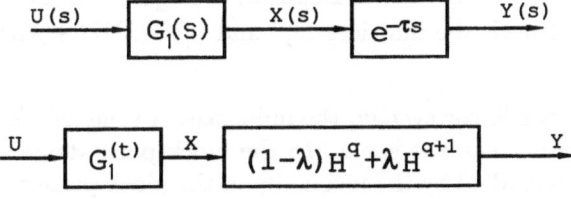

Case 2. Output signal contains time delay

Figure 6.9: Block diagram of a system containing time delay, $\tau = (q + \lambda)h$.

we have:

$$Y^T = U^T \left((1-\lambda)H^q + \lambda H^{q+1} \right) G_1^{(t)} \tag{6.70}$$

Equation (6.70) shows that the block pulse transfer matrix of a system containing time delay in the input side is:

$$G^{(t)} = \left((1-\lambda)H^q + \lambda H^{q+1} \right) G_1^{(t)} \tag{6.71}$$

For the case of time delay in the output side, from the relations:

$$X^T = U^T G_1^{(t)} \tag{6.72}$$

and

$$Y^T = X^T \left((1-\lambda)H^q + \lambda H^{q+1} \right) \tag{6.73}$$

we have:

$$Y^T = U^T G_1^{(t)} \left((1-\lambda)H^q + \lambda H^{q+1} \right) \tag{6.74}$$

Equation (6.74) shows that the block pulse transfer matrix of a system containing time delay in the output side is:

$$G^{(t)} = G_1^{(t)} \left((1-\lambda)H^q + \lambda H^{q+1} \right) \tag{6.75}$$

In both cases of time delay in the input side and in the output side, a constant matrix is introduced in the corresponding block pulse transfer matrices.

For a time-invariant linear system, the time delay in the input side or in the output side has the same influence. In the Laplace domain, this same influence is embodied by the interchangeability of the two factors in the system transfer function:

$$e^{-\tau s} G_1(s) = G_1(s) e^{-\tau s} \tag{6.76}$$

In the block pulse domain, this same influence is also embodied by the interchangeability of the two matrices in the block pulse transfer matrix of the system:

$$\begin{aligned} G^{(t)} &= \left((1-\lambda)H^q + \lambda H^{q+1} \right) G_1^{(t)} \\ &= G_1^{(t)} \left((1-\lambda)H^q + \lambda H^{q+1} \right) \end{aligned} \tag{6.77}$$

The interchangeability of matrix multiplication in (6.77) is due to the special upper triangular forms of both the matrices $G_1^{(t)}$ and $((1-\lambda)H^q + \lambda H^{q+1})$, as concluded in Appendix B.

For a time-varying linear system, the influences of time delays in the input side and in the output side are not the same. In the block pulse domain, these two different influences can be embodied by the inequality of the block pulse transfer matrices of the both cases:

$$\left((1-\lambda)H^q + \lambda H^{q+1} \right) G_1^{(t)} \neq G_1^{(t)} \left((1-\lambda)H^q + \lambda H^{q+1} \right) \tag{6.78}$$

In fact, this inequality can be traced back to $G_1^{(t)}$, which has no more the special upper triangular form, also as concluded in Appendix B.

The above discussions about some elementary linear systems give principles of using block pulse transfer matrices in the block diagrams. These principles are similar to the uses of Laplace transfer functions. But block pulse transfer matrices are more flexible, because they can easily be applied in both time-invariant and time-varying cases. The only thing we should take care of is that no interchangeability of matrices can be used any more in the construction of block pulse transfer matrices of whole systems if some subsystems are time-varying. Here is an example to show the principles of constructing block pulse transfer matrices in multiloop cases.

Example 6.3 Determine the block pulse transfer matrix of a multiloop system from the block diagram in Figure 6.10.

In the block pulse domain, the following relations can be established from the block diagram of Figure 6.10:

$$X_1^T = U^T + X_3^T + X_4^T$$
$$X_2^T = X_1^T G_1^{(t)}$$
$$X_3^T = X_2^T G_3^{(t)}$$
$$X_4^T = Y^T G_4^{(t)}$$
$$Y^T = X_2^T G_2^{(t)}$$

From these relations, the response of the system can be obtained:

$$Y^T = U^T G_1^{(t)} G_2^{(t)} \left(I - \left(G_2^{(t)} \right)^{-1} G_3^{(t)} G_1^{(t)} G_2^{(t)} - G_4^{(t)} G_1^{(t)} G_2^{(t)} \right)^{-1} \qquad (6.79)$$

Therefore the block pulse transfer matrix of this multiloop system is:

$$G^{(t)} = G_1^{(t)} G_2^{(t)} \left(I - \left(G_2^{(t)} \right)^{-1} G_3^{(t)} G_1^{(t)} G_2^{(t)} - G_4^{(t)} G_1^{(t)} G_2^{(t)} \right)^{-1}$$

which describes the dynamics of this system in a piecewise constant approximate way, regardless whether its parts are time-varying or not. But if this multiloop system is time-invariant, its block pulse transfer matrix can further be simplified to:

$$G^{(t)} = \left(I - G_1^{(t)} G_3^{(t)} - G_1^{(t)} G_2^{(t)} G_4^{(t)} \right)^{-1} G_1^{(t)} G_2^{(t)}$$

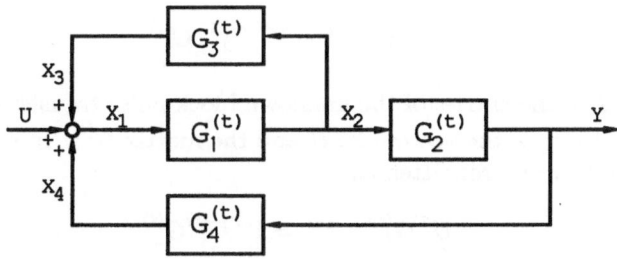

Figure 6.10: Block diagram of a multiloop system.

because the order of the matrices $G_1^{(t)}$, $G_2^{(t)}$, $G_3^{(t)}$, $G_4^{(t)}$ can now be changed in the multiplication.

6.4   Identification of block pulse transfer matrices

In the block pulse domain, the problem of identifying nonparameteric models is to estimate the block pulse transfer matrices of continuous-time linear systems based on the measured data of input and output signals. Like the impulse responses of systems, no prior knowledge of system order is required in this nonparameteric model identification problem, therefore the block pulse transfer matrices are suitable for the exploratory studies of linear systems.

1. Identification of time-invariant linear systems. If all the parts of linear systems are time-invariant, the corresponding block pulse transfer matrices have special upper triangular forms. In these cases, only m entries of the unknown block pulse transfer matrix should be determined in our identification problem. Here are the identifications of two simple systems.

When the input and output signals of a part of a system can be measured directly, the block pulse transfer matrix of this part can be determined at once. From the block diagram in Figure 6.5, the relation between the input and output of this part can be expressed as:

$$Y^T = U^T G^{(t)} \tag{6.80}$$

According to (B.12) in Appendix B, the entries of the unknown block pulse transfer matrix $G^{(t)}$ can be determined from those of the vectors U and Y.

When only the input and output signals of a system can be measured, in which two or more subsystems are included, the block pulse transfer matrix of one subsystem can be determined without breaking the loops if the block pulse transfer matrices of other subsystems are known. From Figure 6.8, the relation between the input and output of the simple feedback system is:

$$Y^T = U^T G_1^{(t)} \left(I + G_2^{(t)} G_1^{(t)} \right)^{-1} \tag{6.81}$$

In case the matrix $G_1^{(t)}$ is unknown, (6.81) can be rewritten as:

$$Y^T = \left(U^T - Y^T G_2^{(t)} \right) G_1^{(t)} \tag{6.82}$$

According to (B.12), the entries of the unknown block pulse transfer matrix $G_1^{(t)}$ can be determined from those of the vectors Y, U and the matrix $G_2^{(t)}$. In case the matrix $G_2^{(t)}$ is unknown, (6.81) can be rewritten as:

$$U^T G_1^{(t)} - Y^T = Y^T G_1^{(t)} G_2^{(t)} \tag{6.83}$$

in which the order of the matrices $G_1^{(t)}$ and $G_2^{(t)}$ is changed in the multiplication due to their special upper triangular forms. According to (B.12), the entries of the unknown block pulse transfer matrix $G_2^{(t)}$ can be determined from those of the vectors Y, U and the matrix $G_1^{(t)}$.

The above discussions about the identification of block pulse transfer matrices in simple systems can also be used in multiloop systems, if the block pulse transfer matrix of one subsystem is unknown. In time-invariant cases, the identification problem is rather simple because the order of all the block pulse transfer matrices can be changed flexibly in multiplications due to their special upper triangular forms. Here is an example.

Example 6.4 Determine the unknown block pulse transfer matrix $G_4^{(t)}$ of a multiloop time-invariant linear system in Figure 6.10 from the input U, output Y and the known block pulse transfer matrices $G_j^{(t)}$ $(j = 1, 2, 3)$ without breaking the loops.

Since the order of the block pulse transfer matrices can be changed in the multiplication, (6.79) can be rewritten as:

$$Y^T - Y^T G_1^{(t)} G_3^{(t)} - U^T G_1^{(t)} G_2^{(t)} = Y^T G_1^{(t)} G_2^{(t)} G_4^{(t)}$$

According to (B.12) in Appendix B, the entries of the unknown $G_4^{(t)}$ can be determined from those of the vectors Y, U and the matrices $G_j^{(t)}$ $(j = 1, 2, 3)$:

$$g_{4,i}^{(t)} = \begin{cases} \dfrac{1}{v_1} w_1 & \text{for } i = 1 \\[2mm] \dfrac{1}{v_1}\left(w_i - \displaystyle\sum_{k=2}^{i} v_k g_{4,i-k+1}^{(t)} \right) & \text{for } i = 2, 3, \ldots, m \end{cases}$$

where v_k, w_k $(k = 1, 2, \ldots, m)$ are entries of the vectors:

$$V^T = U^T G_1^{(t)} G_2^{(t)}$$

and

$$W^T = Y^T - Y^T G_1^{(t)} G_3^{(t)} - U^T G_1^{(t)} G_2^{(t)}$$

Obviously, the special upper triangular forms of all the block pulse transfer matrices $G_j^{(t)}$ $(j = 1, \ldots, 4)$ simplify the identification procedure.

2. Identification of time-varying linear systems. Since the corresponding block pulse transfer matrices of time-varying linear systems have general upper triangular forms, $m(m+1)/2$ entries of the unknown block pulse transfer matrix should be determined. Usually, we use data of m independent inputs and their corresponding outputs to solve our identification problem. Here are the identifications of two simple systems.

Similar to the time-invariant case, when the input and output signals of a subsystem can be measured directly, the block pulse transfer matrix of this subsystem can easily be determined. If the block pulse coefficient vectors of m sets of input and output signals are expressed in matrix forms:

$$\bar{U} = \begin{pmatrix} U_1 & U_2 & \cdots & U_m \end{pmatrix} \tag{6.84}$$

and

$$\bar{Y} = \left(\begin{array}{cccc} Y_1 & Y_2 & \cdots & Y_m \end{array}\right) \tag{6.85}$$

(6.80) gives the relation:

$$\bar{Y}^T = \bar{U}^T G^{(t)} \tag{6.86}$$

Since the matrix $\bar{U}$ is usually nonsingular due to the independence of the input signals, the block pulse transfer matrix $G^{(t)}$ can be obtained as:

$$G^{(t)} = \left(\bar{U}^T\right)^{-1} \bar{Y}^T \tag{6.87}$$

Similar to the time-invariant case, when only the input and output signals of a system can be measured, in which two or more subsystems are included, the block pulse transfer matrix of one subsystem can be determined without breaking the loops if the block pulse transfer matrices of other subsystems are known. From Figure 6.8, the relation between the m sets of input and output signals of the simple feedback system is:

$$\bar{Y}^T = \bar{U}^T G_1^{(t)} \left(I + G_2^{(t)} G_1^{(t)}\right)^{-1} \tag{6.88}$$

In case the matrix $G_1^{(t)}$ is unknown, (6.88) can be rewritten as:

$$\bar{Y}^T = \left(\bar{U}^T - \bar{Y}^T G_2^{(t)}\right) G_1^{(t)} \tag{6.89}$$

therefore the unknown matrix $G_1^{(t)}$ can be determined from those of the matrices $\bar{Y}$, $\bar{U}$ and $G_2^{(t)}$:

$$G_1^{(t)} = \left(\bar{U}^T - \bar{Y}^T G_2^{(t)}\right)^{-1} \bar{Y}^T \tag{6.90}$$

In case the matrix $G_2^{(t)}$ is unknown, (6.88) can be rewritten as:

$$\bar{U}^T G_1^{(t)} - \bar{Y}^T = \bar{Y}^T G_2^{(t)} G_1^{(t)} \tag{6.91}$$

therefore the unknown matrix $G_2^{(t)}$ can be determined from those of the matrices $\bar{Y}$, $\bar{U}$ and $G_1^{(t)}$:

$$G_2^{(t)} = \left(\bar{Y}^T\right)^{-1} \left(\bar{U}^T G_1^{(t)} - \bar{Y}^T\right) \left(G_1^{(t)}\right)^{-1} \tag{6.92}$$

Comparing with the solution of (6.83), the solution of (6.92) is more complicated, because the order of the block pulse transfer matrices with general upper triangular forms can not be changed with each other in multiplications.

Special input signals can be used to simplify the determination of the block pulse transfer matrices in the identification of time-varying linear systems. If the m input signals have the same forms but with different time delays:

$$u_j(t) = f(t - (j-1)h)\mu(t - (j-1)h) \tag{6.93}$$

where $j = 1, 2, \ldots, m$, $\bar{U}^T$ and $\bar{Y}^T$ are upper triangular matrices. In such cases, the entries of the unknown block pulse transfer matrices can be determined from (B.6) and

(B.7) in Appendix B to avoid the inversion operation. For example, the entries of the matrix $G^{(t)}$ in (6.86) can be determined by:

$$
g_{i,j}^{(t)} = \begin{cases} 0 & \text{for } j < i \\[2mm] \dfrac{1}{\bar{u}_{i,i}} \bar{y}_{j,i} & \text{for } j = i \\[3mm] \dfrac{1}{\bar{u}_{i,i}} \left(\bar{y}_{j,i} - \sum_{k=i+1}^{j} \bar{u}_{k,i} g_{k,j}^{(t)} \right) & \text{for } j > i \end{cases}
\tag{6.94}
$$

If the m input signals are chosen more specially as the m block pulse functions:

$$
u_j(t) = \phi_j(t)
\tag{6.95}
$$

the above determination of the unknown block pulse transfer matrices can be simplified further, because the matrix $\bar{U}$ is now reduced to a unit matrix. For example, the matrix $G^{(t)}$ in (6.86) can be obtained just from:

$$
G^{(t)} = \bar{Y}^T
\tag{6.96}
$$

The above discussions about the identification of block pulse transfer matrices in simple systems can also be used in multiloop systems, if the block pulse transfer matrix of one subsystem is unknown. But in time-varying cases, the identification problem becomes complicated, because the order of block pulse transfer matrices can no more be changed in multiplications. Algebraic equation sets should be solved to determine the entries of the unknown block pulse transfer matrix. In order to obtain a better approximation of the system dynamics, the number of the block pulse functions m must be large. But a larger value of m causes a larger size of computation. From this point of view, the block pulse transfer matrices are suitable only for the exploratory studies of linear systems.

Chapter 7

Input-output representations of dynamic systems

In the time domain, dynamic characteristics of continuous-time systems are usually described by differential equations which are represented in the input-output and state space forms. Based on such representations, problems of control and systems science can be studied in detail. But in order to express the relations of variables in these problems more directly, analytical solutions of differential equations should be found, and this is not always an easy task. In the block pulse domain, the original problems are approximated, but the expressions for obtaining their numerical solutions are more direct due to the simple operation rules of block pulse functions. In this chapter, some block pulse function methods are discussed in the input-output representations. The block pulse function methods in the state space representations will be discussed in the next chapter.

7.1 Single-input single-output time-invariant linear systems

As a basic and simple case, we consider first a single-input single-output (SISO) time-invariant linear system which is described by the differential equation:

$$\sum_{i=0}^{n} a_i y^{(i)}(t) = \sum_{i=0}^{n} b_i u^{(i)}(t) \tag{7.1}$$

where $u(t)$ is the scalar input, $y(t)$ is the scalar output, a_i and b_i $(i = 0, 1, \ldots, n)$ are system parameters. For the convenience of expressions, we use here the general form of the differential equation in which all the terms exist, although some terms may not appear in a real dynamic system.

7.1.1   System analysis

In the block pulse domain, the analysis problem of a continuous time-invariant linear system is to evaluate the block pulse coefficients of the output signal $y(t)$ in a finite interval $t \in [0, T)$ from the system differential equation and the initial values under the input excitation $u(t)$. This problem is similar to the simulation of dynamic systems, but here we obtain only the piecewise constant approximation of the simulated signal.

For a SISO time-invariant linear system, we first integrate the differential equation (7.1) n times successively from 0 to t on both sides:

$$
\sum_{i=0}^{n} a_{n-i} \underbrace{\int_0^t \cdots \int_0^t}_{i \text{ times}} y(t)\, dt \cdots dt - \sum_{i=0}^{n-1} \left(y_0^{(i)} \sum_{j=i}^{n-1} a_{n+i-j} \underbrace{\int_0^t \cdots \int_0^t}_{j \text{ times}} dt \cdots dt \right)
$$

$$
= \sum_{i=0}^{n} b_{n-i} \underbrace{\int_0^t \cdots \int_0^t}_{i \text{ times}} u(t)\, dt \cdots dt - \sum_{i=0}^{n-1} \left(u_0^{(i)} \sum_{j=i}^{n-1} b_{n+i-j} \underbrace{\int_0^t \cdots \int_0^t}_{j \text{ times}} dt \cdots dt \right) \quad (7.2)
$$

where $y_0^{(i)}$ and $u_0^{(i)}$ ($i = 0, 1, \ldots, n-1$) are initial values of $y(t)$ and $u(t)$, respectively. After expanding the input and output signals into their block pulse series:

$$
\begin{aligned}
u(t) &\doteq \begin{pmatrix} u_1 & u_2 & \cdots & u_m \end{pmatrix} \Phi(t) \\
&= U^T \Phi(t)
\end{aligned} \quad (7.3)
$$

and

$$
\begin{aligned}
y(t) &\doteq \begin{pmatrix} y_1 & y_2 & \cdots & y_m \end{pmatrix} \Phi(t) \\
&= Y^T \Phi(t)
\end{aligned} \quad (7.4)
$$

we obtain the block pulse series of the above equation in a vector form:

$$
\left(Y^T \sum_{i=0}^{n} a_{n-i} P_i - E^T \sum_{i=0}^{n-1} \left(y_0^{(i)} \sum_{j=i}^{n-1} a_{n+i-j} P_j \right) \right) \Phi(t)
$$

$$
\doteq \left(U^T \sum_{i=0}^{n} b_{n-i} P_i - E^T \sum_{i=0}^{n-1} \left(u_0^{(i)} \sum_{j=i}^{n-1} b_{n+i-j} P_j \right) \right) \Phi(t) \quad (7.5)
$$

Instead of the powers of the conventional integration operational matrix P^j, we use here the generalized integration operatoral matrices P_j for better approximations in the results.

From (7.5), a linear algebraic relation between the block pulse coefficients of the input and output signals can be obtained:

$$
Y^T \sum_{i=0}^{n} a_{n-i} P_i - E^T \sum_{i=0}^{n-1} \left(y_0^{(i)} \sum_{j=i}^{n-1} a_{n+i-j} P_j \right)
$$

$$
\doteq U^T \sum_{i=0}^{n} b_{n-i} P_i - E^T \sum_{i=0}^{n-1} \left(u_0^{(i)} \sum_{j=i}^{n-1} b_{n+i-j} P_j \right) \quad (7.6)
$$

Since the matrix:

$$A = \sum_{i=0}^{n} a_{n-i} P_i \qquad (7.7)$$

is usually nonsigular, the block pulse coefficients of the output can be determined directly from those of the input:

$$Y^T = BA^{-1} \qquad (7.8)$$

where B is a vector:

$$B = U^T \sum_{i=0}^{n} b_{n-i} P_i - E^T \sum_{i=0}^{n-1} \left(u_0^{(i)} \sum_{j=i}^{n-1} b_{n+i-j} P_j \right) + E^T \sum_{i=0}^{n-1} \left(y_0^{(i)} \sum_{j=i}^{n-1} a_{n+i-j} P_j \right) \qquad (7.9)$$

Example 7.1 Determine the block pulse coefficients of the output signal of a second-order linear system:

$$G(s) = \frac{b_0}{s^2 + a_1 s + a_0} \qquad (7.10)$$

in the interval $t \in [0, 10)$ with $m = 20$. Here, the parameters are $a_0 = 2.0$, $a_1 = 3.0$, $a_2 = 1.0$, $b_0 = 1.0$, the initial values are zero, and the input signal is a unit step function.

Since the input signal is a unit step function, i.e. $U = E$, we obtain from (7.8):

$$\begin{aligned} Y^T &= b_0 E^T P_2 \left(a_2 P_0 + a_1 P_1 + a_0 P_2 \right)^{-1} \\ &= \left(\begin{array}{cccc} 0.0227 & 0.1343 & 0.2543 & 0.3431 & \cdots \end{array} \right) \end{aligned} \qquad (7.11)$$

This piecewise constant approximation together with the analytical solution of the output are illustrated in Figure 7.1 for comparison.

7.1.2　System identification

In the block pulse domain, the identification problem of a continuous time-invariant linear system is to estimate the system parameters a_i and b_i $(i = 0, 1, \ldots, n)$ from the

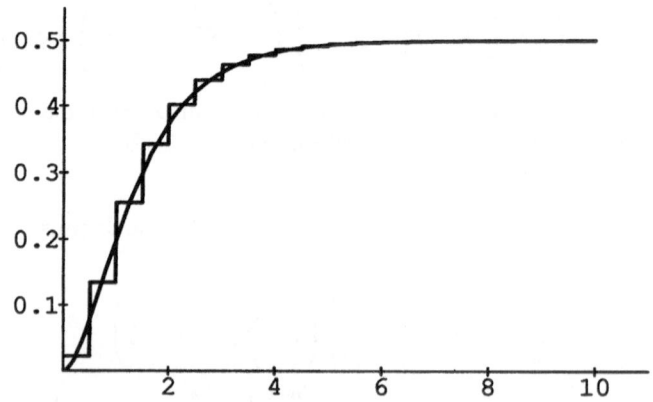

Figure 7.1: System response of a time-invariant linear system.

block pulse coefficients of the input and output signals. This problem is similar to the system identification from the sampled data of the input and output signals, but here we use only the data of the piecewise constant approximation of the continuous signals.

For a SISO time-invariant linear system (7.1), we can set $a_n = 1$ without loss of generality. If all the initial values are unknown, (7.6) can be rewritten as:

$$Y^T = -\sum_{i=1}^{n} a_{n-i} Y^T P_i + \sum_{i=0}^{n} b_{n-i} U^T P_i + \sum_{i=1}^{n} c_i E^T P_{n-i} \tag{7.12}$$

where c_i $(i = 1, 2, \ldots, n)$ are auxiliary parameters:

$$c_i = \sum_{j=i}^{n} \left(a_j y_0^{(j-i)} - b_j u_0^{(j-i)} \right) \tag{7.13}$$

From the linear algebraic equation set (7.12), the parameters of the original differential equation a_0, ..., a_{n-1}, b_0 ..., b_n together with the auxiliary parameters c_1, ..., c_n can be determined, so long as there are more equations than the unknowns. For example, after introducing the vector:

$$\theta^T = \begin{pmatrix} a_0 & \cdots & a_{n-1} & b_0 & \cdots & b_n & c_1 & \cdots & c_n \end{pmatrix} \tag{7.14}$$

and the matrix:

$$G^T = \begin{pmatrix} -Y^T P_n \\ \vdots \\ -Y^T P_1 \\ U^T P_n \\ \vdots \\ U^T P_0 \\ E^T P_{n-1} \\ \vdots \\ E^T P_0 \end{pmatrix} \tag{7.15}$$

all the m equations in (7.12) can be expressed in a matrix form:

$$Y = G\theta \tag{7.16}$$

and the ordinary least-squares algorithm gives:

$$\hat{\theta} = \left(G^T G \right)^{-1} G^T Y \tag{7.17}$$

where $\hat{\theta}$ is the estimated value of the parameter vector θ. This least-squares estimation can also be carried out recursively. After denoting the lth row of the matrix G as the column vector g_l, the parameter vector θ can be estimated by the recursive least-squares algorithm with $l = 1, 2, \ldots, m$:

$$\begin{cases} \hat{\theta}_l = \hat{\theta}_{l-1} + \gamma_l S_{l-1} g_l \left(y_l - g_l^T \hat{\theta}_{l-1} \right) \\ S_l = \dfrac{1}{\alpha} \left(S_{l-1} - \gamma_l S_{l-1} g_l g_l^T S_{l-1} \right) \\ \gamma_l = \dfrac{1}{\alpha + g_l^T S_{l-1} g_l} \end{cases} \tag{7.18}$$

where $\hat{\theta}_{l-1}$ and $\hat{\theta}_l$ are the $(l-1)$th and lth estimated values of θ respectively, $0 < \alpha \leq 1$ is the forgetting factor which has the role of smoothing the past data exponentially. At the beginning of the estimation, we can set $\hat{\theta}_0 = 0$, $S_0 = c^2 I$, with a sufficiently large value of c.

Since the linear algebraic equation set (7.12) contains the same parameters as the original differential equation (7.1) in which all the parameters a_i $(i = 0, 1, \ldots, n-1)$ and b_j $(j = 0, 1, \ldots, n)$ exist, a simpler form can be obtained directly from some a priori knowledge about the system. If some terms do not exist in the original differential equation, the corresponding terms in the linear algebraic equation set can be eliminated. If some parameters or initial values of the original differential equation are known a priori from their physical meanings, the corresponding terms in the linear algebraic equation set can be included in its left hand side. In these cases, the dimensions of the above defined vectors and matrices can be reduced, and the estimation process can be simplified.

To construct the linear algebraic equation set (7.12) for the identification problem, a series of matrix operations should be done, i.e. for the input and output signals, all the block pulse coefficients of their one time to n times integrals should be computed separately based on the integration operational matrices. Due to these operations, the size of computation is large, especially when the number of block pulse functions m is large. In order to reduce the computations involved, the recursive formula for evaluating block pulse coefficients of multiple integrals (5.69) is useful here. In using this recursive formula, the block pulse coefficients of the corresponding multiple integrals are not computed separately. In contrast, when the block pulse coefficients of the n times integrals of a signal is evaluated from this recursive formula, the intermediate results in the computation are just the block pulse coefficients of the one time to $(n-1)$ times integrals of the same signal. Since there are no unnecessary intermediate results in the computation, this recursive formula can be used here efficiently to reduce the size of computation in the data preparation stage of system identification.

Example 7.2 Consider a second-order time-invariant linear system described by (7.10) with parameters $a_0 = 2.0$, $a_1 = 3.0$ and $b_0 = 1.0$. The simulated output of the system is obtained from the input $u(t) = \sin t + \sin 2t$. Our problem is to estimate the unknown system parameters from the given input and output data. Here the sampling period is $h = 0.05$.

According to (7.12), the algebraic equation set becomes:

$$Y^T = -a_1 Y^T P_1 - a_0 Y^T P_2 + b_0 U^T P_2 + c_2 E^T P_0 + c_1 E^T P_1$$

To estimate the system parameters from this equation set, the block pulse coefficients of the input and output signals should be known. When only sampled data of signals are available, these block pulse coefficients can be obtained from various interpolation formulas. As a practical variation, we use here and in other identification examples of this book the two points approximation formula (1.17). In Chapter 9, we will also discuss the three points approximation formula for block pulse coefficients. After the approximate block pulse coefficients of the input and output signals are evaluated, the ordinary least-squares algorithm (7.17) gives the estimation results. With $m = 200$, they are $a_0 = 2.0015$, $a_1 = 3.0034$ and $b_0 = 1.0011$. Discussions about block pulse function methods for solving analysis and identification problems of time-invariant linear systems can be found in the papers of Palanisamy and Bhattacharya (1981b), Cheng and Hsu (1982b).

7.1.3 Block pulse regression equations

As shown in the above discussions, the transformation of original differential equations into their corresponding block pulse expressions is rather straightforward. But there are also disadvantages in these block pulse function methods. As an example, we mention here only the disadvantages in the identification case. Firstly, since the data preparation of (7.12) involves the matrix products $Y^T P_k$, $U^T P_k$ and $E^T P_k$ $(k = 1, 2, \ldots, n)$, and since the operational matrices P_k have upper triangular forms, the evaluation of every single entry in these matrix products must always begin from the first entry of the vectors Y, U and E. The higher the position of an entry of these matrix products, the larger the computations that must be done. This difficulty becomes severe especially when a large amount of data is involved in the identification problem. Clearly, the block pulse function method discussed above is not suitable for practical on-line identification, although recursive algorithms can be applied in the stage of parameter estimation. Secondly, the initial values must either be known a priori or be estimated together with the system parameters. In many cases, we do not care about these initial values, but their existence complicates the estimation procedures and increases the size of computation. Besides, both the way of off-line estimation and the existence of initial values hinder the estimations to follow the slow changes of parameters, which usually happen in processes running over long time. Due to these disadvantages, the block pulse function method for identification should be improved

Based on the disjointness of block pulse functions, m equations of block pulse coefficients can be obtained from equating each $\phi_l(t)$ $(l = 1, 2, \ldots, m)$ of (7.5) on both sides separately. If we denote the lth equation as $E_{(l)}$, it has the form:

$$
\begin{aligned}
\sum_{i=0}^{n} & \left(\frac{h^i}{(i+1)!} a_{n-i} \sum_{j=1}^{l} y_j p_{i,l+1-j} \right) \\
& - \sum_{i=0}^{n-1} \left(y_0^{(i)} \sum_{j=i}^{n-1} \left(\frac{h^j}{(j+1)!} a_{n+i-j} \sum_{c=1}^{l} p_{j,l+1-j-c} \right) \right) \\
= & \sum_{i=0}^{n} \left(\frac{h^i}{(i+1)!} b_{n-i} \sum_{j=1}^{l} u_j p_{i,l+1-j} \right) \\
& - \sum_{i=0}^{n-1} \left(u_0^{(i)} \sum_{j=i}^{n-1} \left(\frac{h^j}{(j+1)!} b_{n+i-j} \sum_{c=1}^{l} p_{j,l+1-j-c} \right) \right)
\end{aligned}
\tag{7.19}
$$

Applying the operation $\sum_{k=0}^{n}(-1)^k \binom{n}{k} E_{(l+n-k)}$ on the $(n+1)$ successive equations $E_{(l)}$, $E_{(l+1)}, \ldots, E_{(l+n)}$ $(1 \le l \le m-n)$, and noticing that all the terms involving initial values are zero according to (5.43), we obtain:

$$\sum_{i=0}^{n} \frac{h^i}{(i+1)!} a_{n-i} Q_i = \sum_{i=0}^{n} \frac{h^i}{(i+1)!} b_{n-i} R_i \qquad (7.20)$$

where

$$
\left.
\begin{aligned}
Q_i = y_1 \sum_{k=0}^{n}(-1)^k \binom{n}{k} p_{i,l+n-k} \\[2mm]
+ y_2 \sum_{k=0}^{n}(-1)^k \binom{n}{k} p_{i,l+n-k-1} \\[2mm]
+ \cdots \\[2mm]
+ y_{l-1} \sum_{k=0}^{n}(-1)^k \binom{n}{k} p_{i,l+n-k-(l-2)}
\end{aligned}
\right\} \text{Part 1}
$$

$$
\left.
+ y_l \sum_{k=0}^{n}(-1)^k \binom{n}{k} p_{i,l+n-k-(l-1)}
\right\} \text{Part 2} \qquad (7.21)
$$

$$
\left.
\begin{aligned}
+ y_{l+1} \sum_{k=0}^{n-1}(-1)^k \binom{n}{k} p_{i,l+n-k-l} \\[2mm]
+ y_{l+2} \sum_{k=0}^{n-2}(-1)^k \binom{n}{k} p_{i,l+n-k-(l+1)} \\[2mm]
+ \cdots \\[2mm]
+ y_{l+n} \sum_{k=0}^{0}(-1)^k \binom{n}{k} p_{i,l+n-k-(l+n-1)}
\end{aligned}
\right\} \text{Part 3}
$$

If we substitute u_j for y_j and R_i for Q_i in the above equations, we can also obtain a similar expression of R_i.

Equation (7.21) can further be simplified by the relations between the entries of the generalized integration operatoral matrices. Part 1 exists only when $l > 1$, and this part equals zero according to (5.37), whereas Part 2 equals $(-1)^{n+i} y_l$ according to (5.41). Therefore (7.20) can be rewritten as:

$$\sum_{k=0}^{n} A_k y_{l+k} = \sum_{k=0}^{n} B_k u_{l+k} \qquad (7.22)$$

where A_j, B_j $(j = 0, 1, \ldots, n)$ are linear combinations of the original parameters a_i, b_i $(i = 0, 1, \ldots, n)$, respectively. Since this equation expresses the relation between the input and output signals of a SISO linear system in a piecewise constant approximate way, it can be regarded as the block pulse difference equation corresponding to the differential equation (7.1). But this difference equation contains no derivatives of signals and no initial values, and the relation between the block pulse coefficients of the input and output signals is much simpler than (7.6).

From the above derivation, we obtain the relation between a_i and A_j:

$$A_j = \sum_{i=0}^{n} \left(\frac{h^i}{(i+1)!} a_{n-i} \sum_{k=0}^{n-j} (-1)^k \binom{n}{k} p_{i,n-k-j+1} \right) \qquad (7.23)$$

If we substitute B_j for A_j and b_i for a_i in the above expression, we can also obtain a similar relation between b_i and B_j. But in fact, the relations between the two kinds of parameters can be expressed more simplily. For example, for the second-order systems, we have:

$$\begin{pmatrix} A_2 \\ A_1 \\ A_0 \end{pmatrix} = \begin{pmatrix} 1 & 1 & 1 \\ -2 & 0 & 4 \\ 1 & -1 & 1 \end{pmatrix} \begin{pmatrix} a_2 \\ \dfrac{h}{2} a_1 \\ \dfrac{h^2}{6} a_0 \end{pmatrix} \qquad (7.24)$$

Here, each column of the coefficient matrix can be determined directly from the second-order difference of the first three entries of the matrices P_0, P_1 and P_2 respectively, as illustrated in Figure 7.2. As a general rule, the relation between a_i and A_j of an nth-order system can be obtained from the nth-order difference of the entries of the generalized integration operatoral matrices. The ith column of the coefficient matrix like the one in (7.24) can be determined from the first $(n+1)$ entries of the generalized integration operatoral matrix $p_{i-1,1}, p_{i-1,2}, \ldots, p_{i-1,n+1}$ as defined in (5.31). For the convenience of use, we list the relations between the parameters a_i and A_j for various order systems in Appendix C.

Based on the block pulse difference equation (7.22), the analysis problem of continuous time-invariant linear systems can easily be solved in the block pulse domain. If the first n block pulse coefficients $y_1, y_2, \ldots, y_n$ are known, the $(n+1)$th block pulse coefficients y_{n+1} can be determined from them. After y_{n+1} is obtained, y_{n+2} can also be evaluated from $y_2, y_3, \ldots, y_{n+1}$. Thus, the recursion can be continued to evaluate all the block pulse coefficients of the output. As the initial values of the block pulse difference equation, the first n block pulse coefficients of the output can be evaluated directly from (7.19) in which the initial values of the original differential equation (7.1) are involved.

Using the block pulse difference equation, the same analysis problem as in Example 7.1 can easily be solved. Since $A_1/A_2 = -0.9091$, $A_0/A_2 = 0.1819$, $B_2/A_2 = 0.0227$, $B_1/A_2 = 0.0909$ and $B_0/A_2 = 0.0227$, the block pulse difference equation (7.22) becomes:

$$y_{l+2} = 0.9091 y_{l+1} - 0.1819 y_l + 0.0227 u_{l+2} + 0.0909 u_{l+1} + 0.0227 u_l \qquad (7.25)$$

The block pulse coefficients of the output can then be evaluated recursively, which are the same as those obtained from the block pulse function method in (7.11).

Although the block pulse function methods are rather straightforward, the accuracies of these methods are not high. For the problem of system simulation, better solutions can be obtained from other numerical methods. Here is an example to show the accuracies of some simulation methods.

Example 7.3 Consider a second-order linear system described by (7.10) under the input excitation $u(t) = \sin(t)$. Compare the simulation results obtained from the block pulse difference equation method, the Euler method and the fourth-order Runge-Kutta method.

Since the width of block pulses is fixed, we use the Euler and Runge-Kutta methods also with a fixed step h. In order to compare the simulated outputs of these methods at the sampling instants, each block pulse coefficient of the input and output signals is approximated by the mean value of the signal at the two end points of the corresponding subinterval. From (1.17) and (7.22), we have:

$$(2h^2 + 9h + 6)\bar{y}_{k+3} + (10h^2 + 9h - 6)\bar{y}_{k+2} + (10h^2 - 9h - 6)\bar{y}_{k+1}$$

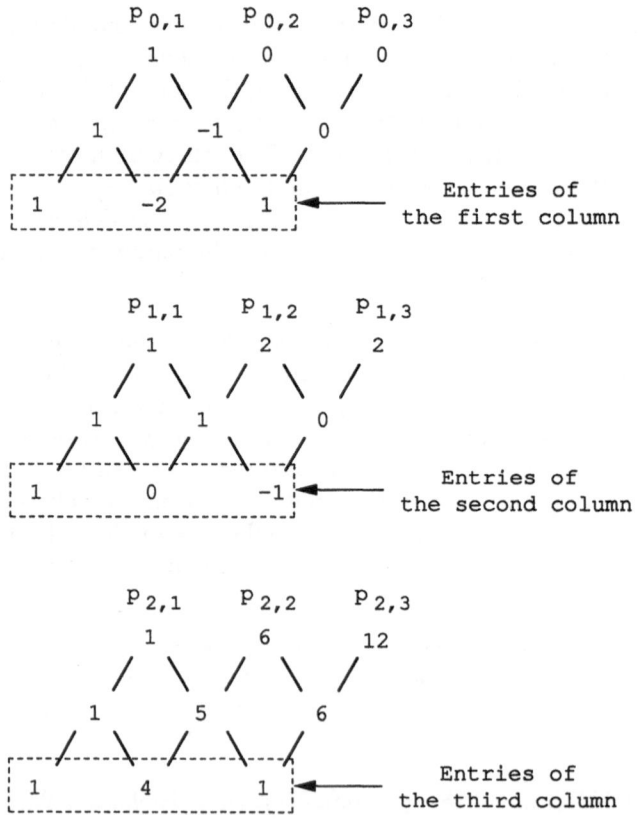

Figure 7.2: Computation of entries in coefficient matrix of (7.24).

$$+(2h^2 - 9h + 6)\bar{y}_k = h^2 \left(\bar{u}_{k+3} + 5\bar{u}_{k+2} + 5\bar{u}_{k+1} + \bar{u}_k\right) \tag{7.26}$$

where $\bar{u}_k$ and $\bar{y}_k$ are the sampled values of the continuous input and output signals at the time instant $t = kh$, respectively. Using this equation, the discrete values of the system response at the sampling instants can be computed by the block pulse difference equation method. Similarly, the discrete values of the system response can also be computed by the Euler and Runge-Kutta methods, respectively. To assess the qualities of these simulation methods, the obtained results are compared with the exact response of the original system:

$$y(t) = \frac{1}{2}e^{-t} - \frac{1}{5}e^{-2t} + \frac{1}{10}\sin(t) - \frac{3}{10}\cos(t) \tag{7.27}$$

Using the first 200 discrete values in each case, the performance index is expressed as:

$$\bar{e}_j = \sum_{k=1}^{200} |\bar{y}_k - \bar{y}_{j,k}| \tag{7.28}$$

where j $(j = 1, 2, 3)$ indicate the different approximation methods, $\bar{y}_k$ and $\bar{y}_{j,k}$ are the discrete values of the exact system response and its jth approximation, respectively. For various sampling periods, these differences are listed in Table 7.1. It is obvious that the method based on the block pulse functions is worse than the Runge-Kutta method and better than the Euler method. Since relations between variables can easily be expressed in the block pulse domain for the problems involving integrations and derivations, the methods based on block pulse functions are feasible in the simulation of systems, especially in the exploratory studies of systems.

Based on the block pulse difference equation (7.22), the identification problem of continuous time-invariant linear systems can also be solved more efficiently. The parameters of the block pulse difference equation A_k and B_k can first be estimated from the block pulse coefficients of the input and output signals, and then be transformed

Sampling period h	0.005	0.05	0.5
Method 1	0.000063	0.003648	0.338513
Method 2	0.053996	1.039596	13.650636
Method 3	0.000002	0.0000087	0.165835

Method 1 — Block pulse difference equation method
Method 2 — Euler method
Method 3 — Runge-Kutta method

Table 7.1: Comparison of different simulation methods.

into the parameters of the original differential equation a_k and b_k, because (7.23) is a linear combination between these two kinds of parameters.

As a more convenient form for the identification problem, we can also obtain a block pulse regression equation after substituting (7.23) into (7.22), which contains the parameters of the original differential equation:

$$z_{n,l} = -\sum_{k=0}^{n-1} a_k z_{k,l} + \sum_{k=0}^{n} b_k v_{k,l} \tag{7.29}$$

where $z_{k,l}$ and $v_{k,l}$ $(k = 0, 1, \ldots, n)$ are linear combinations of the block pulse coefficients y_{l+j} and u_{l+j} $(j = 0, 1, \ldots, n)$, respectively. The block pulse coefficients y_{l+j} and $z_{k,l}$ are connected by:

$$z_{k,l} = \frac{h^{n-k}}{(n-k+1)!} \sum_{j=0}^{n} \sum_{i=0}^{n-j} (-1)^i \binom{n}{i} p_{n-k,n-i-j+1} y_{l+j} \tag{7.30}$$

and the block pulse coefficients u_{l+j} and $v_{k,l}$ also have a similar relation if we substitute $v_{k,l}$ for $z_{k,l}$ and u_{l+j} for y_{l+j} in (7.30). In this equation form, the relation between the block pulse coefficients y_{l+j} and $z_{k,l}$ for an nth-order system can also be established using the rule of the nth-order difference mentioned above, so that the complicated computations of (7.30) can be avoided. For example, for second-order systems, this relation is:

$$\begin{pmatrix} z_{2,l} \\ \dfrac{2}{h} z_{1,l} \\ \dfrac{6}{h^2} z_{0,l} \end{pmatrix} = \begin{pmatrix} 1 & -2 & 1 \\ 1 & 0 & -1 \\ 1 & 4 & 1 \end{pmatrix} \begin{pmatrix} y_{l+2} \\ y_{l+1} \\ y_l \end{pmatrix} \tag{7.31}$$

Obviously, the coefficient matrix in (7.31) is the transpose matrix of the one in (7.24). For the convenience of use, we also list the relations between the block pulse coefficients y_{l+j} and $z_{k,l}$ for various order systems in Appendix C.

For the system identification problem, the disadvantages of the block pulse function method mentioned above are avoided by the block pulse regression equation method. In the cases of recursive estimations, only a small amount of computations must be done in each step, regardless whether the volume of data involved in the estimation problems is small or large. The elimination of the initial values in the block pulse regression equations also reduces the number of unknowns, so that the computation in the estimation procedures is reduced. Moreover, the forgetting factor enables now the estimations to follow slow changes of parameters.

Example 7.4 Consider the same identification problem as in Example 7.2, only the parameter a_0 is now time-varying:

$$a_0 = \begin{cases} 2.0 & \text{for } t < 15 \\ 2.5 & \text{for } 15 \le t < 35 \\ 2.0 & \text{for } t \ge 35 \end{cases}$$

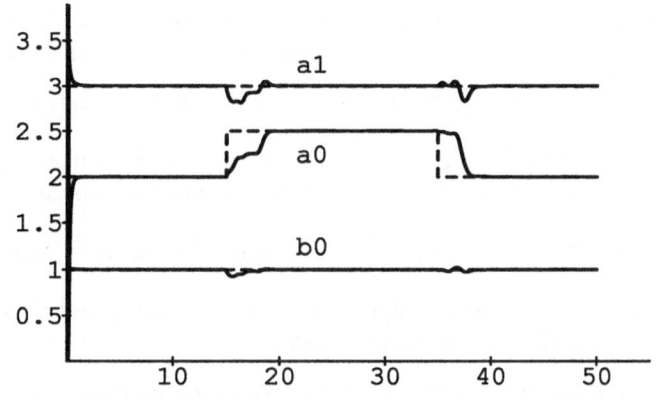

— : estimated parameters,   - - - : true parameters

Figure 7.3: Estimations tracing the changes of the system parameters.

From (7.29), we have the regression equation:

$$y_{l+2} - 2y_{l+1} + y_l = -\frac{h}{2}(y_{l+2} - y_l)a_1$$
$$-\frac{h^2}{6}(y_{l+2} + 4y_{l+1} + y_l)a_0 + \frac{h^2}{6}(u_{l+2} + 4u_{l+1} + u_l)b_0 \qquad (7.32)$$

Setting the exponential forgetting factor $\alpha = 0.95$, the recursive least-squares estimation algorithm gives the estimations of a_0, a_1 and b_0 which can follow the changes of the actual parameters in the system, as illustrated in Figure 7.3. Discussions about block pulse regression equations of time-invariant linear systems can be found in the papers of Jiang and Schaufelberger (1985a,c, 1991a), Kraus and Schaufelberger (1990), Jiang (1990).

7.1.4 Sensitivity analysis

For a time-invariant linear system described by the differential equation:

$$a_n y^{(n)}(t) + a_{n-1} y^{(n-1)}(t) + \cdots + a_1 y^{(1)}(t) + a_0 y(t) = u(t) \qquad (7.33)$$

the output sensitivity function related to the parameter a_j is defined as:

$$\sigma_j = \sigma(t, a_j) = \frac{\partial y}{\partial a_j} \qquad (7.34)$$

Using the output sensitivity function σ_j, the deviation of the output Δy from its nominal output y_0 due to the deviation Δa_j of the parameter a_j from its nominal value $a_{j,0}$ can be described by:

$$\Delta y = \sigma(t, a_{j,0}) \Delta a_j \qquad (7.35)$$

Usually, the output sensitivity function σ_j can be obtained from the following two steps (Frank, 1978). First, the function $y_0^{(j)}$, which is the jth-order derivative of the nominal output, is determined from solving the nominal original differential equation:

$$a_{n,0}y^{(n)}(t) + a_{n-1,0}y^{(n-1)}(t) + \cdots + a_{1,0}y^{(1)}(t) + a_{0,0}y(t) = u(t) \qquad (7.36)$$

with initial values $y^{(k)}(0)$ $(k = 0, 1, \ldots, n-1)$. After that, the output sensitivity function σ_j is determined from solving the nominal output sensitivity equation:

$$a_{n,0}\sigma_j^{(n)} + a_{n-1,0}\sigma_j^{(n-1)} + \cdots + a_{1,0}\sigma_j^{(1)} + a_{0,0}\sigma_j = -y_0^{(j)} \qquad (7.37)$$

with initial values $\sigma_j^{(k)}(0) = 0$ $(k = 0, 1, \ldots, n-1)$. In (7.36) and (7.37), $a_{k,0}$ $(k = 0, 1, \ldots, n-1)$ are the nominal values of parameters a_k.

Since both of these steps are about the solutions of differential equations, their expressions in the block pulse domain can be obtained rather straightforwardly. After integrating (7.36) and (7.37) n times successively from 0 to t on both sides, we can first evaluate the block pulse coefficients of $y(t)$ from:

$$Y^T \sum_{i=0}^{n} a_{n-i,0}P_i - E^T \sum_{i=0}^{n-1} \left(y_0^{(i)} \sum_{j=i}^{n-1} a_{n+i-j,0}P_j \right) = U^T P_n \qquad (7.38)$$

and then evaluate the block pulse coefficients of σ_j from:

$$\Sigma_j^T \sum_{i=0}^{n} a_{n-i,0}P_i = -Y^T P_{n-j} \qquad (7.39)$$

Both (7.38) and (7.39) are the variations of (7.5). Here, the block pulse difference equation (7.22) and the recursive formula for multiple integrals (5.69) can also be used to compute the result recursively. This recursion reduces the size of computation and makes the procedure for output sensitivity analysis problem more efficient.

Example 7.5 Determine the block pulse coefficients of the output sensitivity functions σ_j $(j = 0, 1, 2)$ of a third-order linear system:

$$y^{(3)}(t) + a_2 y^{(2)}(t) + a_1 y^{(1)}(t) + a_0 y(t) = u(t)$$

in the interval $t \in [0, 6)$ with $h = 0.2$. The nominal values of parameters are $a_{2,0} = 5$, $a_{1,0} = 15$, $a_{0,0} = 20$, the input signal is a unit step function and the initial values are zero.

If we denote the sum of the matrices as:

$$M = P_0 + a_{2,0}P_1 + a_{1,0}P_2 + a_{0,0}P_3$$

the block pulse coefficients of the output sensitivity functions can be determined from (7.38) and (7.39) as:

$$\begin{aligned}
\Sigma_0^T &= -U^T P_3 M^{-1} P_3 M^{-1} \\
&= 10^3 \times \left(\begin{array}{cccc} -0.0000 & -0.0012 & -0.0114 & -0.0545 \end{array} \cdots \right)
\end{aligned}$$

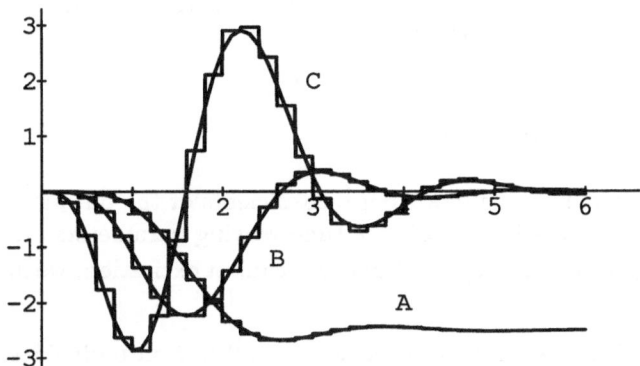

Figure 7.4: Output sensitivity functions of a time-invariant linear system.

$$\begin{aligned} \Sigma_1^T &= -U^T P_3 M^{-1} P_2 M^{-1} \\ &= 10^3 \times \left(\begin{array}{cccc} -0.0000 & -0.0163 & -0.1058 & -0.3578 & \cdots \end{array} \right) \end{aligned}$$

and

$$\begin{aligned} \Sigma_2^T &= -U^T P_3 M^{-1} P_1 M^{-1} \\ &= 10^3 \times \left(\begin{array}{cccc} -0.0129 & -0.1923 & -0.7921 & -1.762 & \cdots \end{array} \right) \end{aligned}$$

The piecewise constant approximations of the output sensitivity functions together with their analytical solutions are illustrated in Figure 7.4, where A, B and C indicate the curves σ_0, σ_1 and σ_2, respectively. Discussions about block pulse function methods for solving sensitivity analysis problem of time-invariant linear systems can be found in the papers of Wang and Marleau (1985), Jiang (1986), Kekkeris and Marszalek (1989).

7.2 Single-input single-output time-varying linear systems

Consider a time-varying linear system which is described by the differential equation:

$$\sum_{k=0}^{N} a_k(t) y^{(k)}(t) = \sum_{k=0}^{N} b_k(t) u^{(k)}(t) \tag{7.40}$$

where $u(t)$ and $y(t)$ are scalar input and output, respectively. The system parameters $a_k(t)$ and $b_k(t)$ are time functions which are described by the nth-order power polynomials with constant coefficients:

$$a_k(t) = a_{k,0} + a_{k,1} t + \ldots + a_{k,n} t^n \tag{7.41}$$

and

$$b_k(t) = b_{k,0} + b_{k,1} t + \ldots + b_{k,n} t^n \tag{7.42}$$

Here for the convenience of expressions, we also use the general form of the differential equation in which all the terms exist.

7.2.1 System analysis

In order to evaluate the block pulse coefficients of the output signal $y(t)$ from the system differential equation (7.40), the time-varying parameters (7.41), (7.42) and the initial values $y_0^{(i)}$ $(i = 0, 1, \ldots, n-1)$ under the input excitation, we first do the following derivation.

Integrating (7.40) N times successively from 0 to t on both sides, we have:

$$\sum_{k=0}^{N} \underbrace{\int_0^t \cdots \int_0^t}_{N \text{ times}} a_k(t) y^{(k)}(t) \, dt \cdots dt = \sum_{k=0}^{N} \underbrace{\int_0^t \cdots \int_0^t}_{N \text{ times}} b_k(t) u^{(k)}(t) \, dt \cdots dt \qquad (7.43)$$

Using the method of integration by parts, each term in the above equation can be written as:

$$\underbrace{\int_0^t \cdots \int_0^t}_{N \text{ times}} a_k(t) y^{(k)}(t) \, dt \cdots dt = \sum_{j=0}^{k} (-1)^j \binom{k}{j} \underbrace{\int_0^t \cdots \int_0^t}_{(N-k+j) \text{ times}} a_k^{(j)}(t) y(t) \, dt \cdots dt$$

$$+ \sum_{s=1}^{k} \sum_{j=0}^{k-s} (-1)^{j+1} \binom{j+s-1}{s-1} a_k^{(j)}(0) y^{(k-s-j)}(0) \underbrace{\int_0^t \cdots \int_0^t}_{(N-s) \text{ times}} dt \cdots dt \qquad (7.44)$$

where $a_k^{(j)}(0)$ and $y^{(k-s-j)}(0)$ are initial values of $a_k^{(j)}(t)$ and $y^{(k-s-j)}(t)$, respectively. Since $a_k^{(j)}(t)$ is also a power polynomial with respect to t:

$$a_k^{(j)}(t) = \left(\begin{array}{cccc} a_{k,0} & a_{k,1} & \cdots & a_{k,n} \end{array} \right) \left(\begin{array}{ccccc} 0 & 0 & \cdots & 0 & 0 \\ 1 & 0 & \cdots & 0 & 0 \\ 0 & 2 & \cdots & 0 & 0 \\ \vdots & \vdots & \ddots & \vdots & \vdots \\ 0 & 0 & \cdots & n & 0 \end{array} \right)^j \left(\begin{array}{c} 1 \\ t \\ \vdots \\ t^n \end{array} \right) \qquad (7.45)$$

the multiple integral of $a_k^{(j)}(t) y(t)$ can be expanded into its block pulse series according to (5.87):

$$\underbrace{\int_0^t \cdots \int_0^t}_{i \text{ times}} a_k^{(j)}(t) y(t) \, dt \cdots dt$$

$$= \left(\begin{array}{cccc} a_{k,0} & a_{k,1} & \cdots & a_{k,n} \end{array} \right) \left(\begin{array}{ccccc} 0 & 0 & \cdots & 0 & 0 \\ 1 & 0 & \cdots & 0 & 0 \\ 0 & 2 & \cdots & 0 & 0 \\ \vdots & \vdots & \ddots & \vdots & \vdots \\ 0 & 0 & \cdots & n & 0 \end{array} \right)^j \left(\begin{array}{c} Y^T P_{i,0} \\ Y^T P_{i,1} \\ \vdots \\ Y^T P_{i,n} \end{array} \right) \Phi(t) \qquad (7.46)$$

From the above equations, the relation between the block pulse coefficients of the input and output signals can be obtained:

$$\sum_{k=0}^{N}\sum_{j=0}^{k}\sum_{s=0}^{n-j}(-1)^j\binom{k}{j}\frac{(j+s)!}{s!}a_{k,j+s}Y^T P_{N-k+j,s}$$

$$+\sum_{k=1}^{N}\sum_{s=1}^{k}\sum_{j=0}^{k-s}(-1)^{j+1}\binom{j+s-1}{s-1}a_k^{(j)}(0)y^{(k-s-j)}(0)E^T P_{N-s,0}$$

$$=\sum_{k=0}^{N}\sum_{j=0}^{k}\sum_{s=0}^{n-j}(-1)^j\binom{k}{j}\frac{(j+s)!}{s!}b_{k,j+s}U^T P_{N-k+j,s}$$

$$+\sum_{k=1}^{N}\sum_{s=1}^{k}\sum_{j=0}^{k-s}(-1)^{j+1}\binom{j+s-1}{s-1}b_k^{(j)}(0)u^{(k-s-j)}(0)E^T P_{N-s,0} \qquad (7.47)$$

Based on this relation, the block pulse coefficients of the system response can be determined from the known system excitation (Hwang and Guo, 1984a).

Example 7.6 Determine the block pulse coefficients of the output signal of a second-order time-varying linear system:

$$(t-1)y^{(2)}(t) - ty^{(1)}(t) + y(t) = u(t)$$

in the interval $t \in [0,1)$ with $m = 10$. The initial values are zero, and the input signal is a unit step function.

According to (7.47), we have:

$$2Y^T P_{2,0} - Y^T P_{1,1} - 2Y^T P_{1,0} + Y^T P_{0,1} - Y^T P_{0,0} = U^T P_{2,0}$$

The block pulse coefficients of the system response can then be obtained from:

$$\begin{aligned} Y^T &= U^T P_{2,0}\left(2P_{2,0} - P_{1,1} - 2P_{1,0} + P_{0,1} - P_{0,0}\right)^{-1} \\ &= \left(\begin{matrix} -0.0016 & -0.0119 & -0.0339 & -0.06866 & \cdots \end{matrix}\right) \end{aligned}$$

This piecewise constant approximate solution is illustrated in Figure 7.5 together with the analytical solution: $y(t) = t - e^t + 1$.

7.2.2 System identification

Our problem here is to estimate the time-varying system parameters $a_k(t)$, $b_k(t)$ ($k = 0, 1, \ldots, N$) from the block pulse coefficients of the input and output signals. Without loss of generality, one parameter can be set as a known value, e.g. $a_{N,0} = 1$. If all the initial values are unknown, (7.47) can be rewritten as:

$$Y^T = -\sum_{k=0}^{N-1}\sum_{j=0}^{k}\sum_{s=0}^{n-j}(-1)^j\binom{k}{j}\frac{(j+s)!}{s!}a_{k,j+s}Y^T P_{N-k+j,s}$$

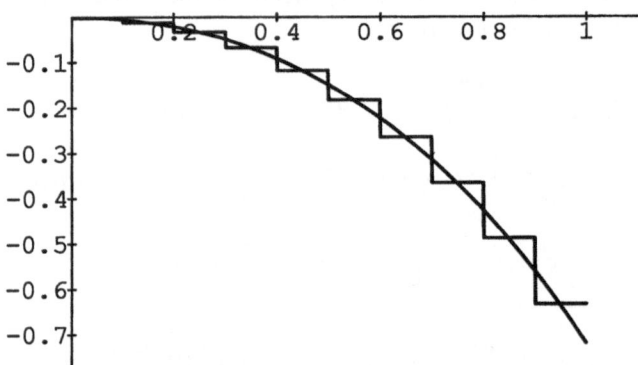

Figure 7.5: System response of a time-varying linear system.

$$-\sum_{j=1}^{N}\sum_{s=0}^{n-j}(-1)^j\binom{N}{j}\frac{(j+s)!}{s!}a_{N,j+s}Y^TP_{j,s}-\sum_{s=1}^{n}a_{N,s}Y^TP_{0,s}$$

$$+\sum_{k=0}^{N}\sum_{j=0}^{k}\sum_{s=0}^{n-j}(-1)^j\binom{k}{j}\frac{(j+s)!}{s!}b_{k,j+s}U^TP_{N-k+j,s}+\sum_{s=1}^{N}c_sE^TP_{N-s,0} \qquad (7.48)$$

where c_s $(s=1,2,\ldots,N)$ are auxiliary parameters:

$$c_s=\sum_{k=s}^{N}\sum_{j=0}^{k-s}(-1)^j\binom{j+s-1}{s-1}\left(a_k^{(j)}(0)y^{(k-s-j)}(0)-b_k^{(j)}(0)u^{(k-s-j)}(0)\right) \qquad (7.49)$$

From the linear algebraic equation set (7.48), the parameters of the original differential equation $a_0(t)$, ..., $a_N(t)$, $b_0(t)$ ..., $b_N(t)$ together with the auxiliary parameters c_1, ..., c_N can be determined, so long as the number of equations is larger than the number of unknowns. In fact, this identification method based on block pulse functions is only an extension of the one discussed in Section 7.1. This can be seen from comparing (7.12) and (7.48) if we set the order of the power polynomials in (7.41) and (7.42) as $n=0$. But also for the same reasons as mentioned before, the large size of computation involved in the matrix operations and the initial values involved in the estimation procedures make the problem of time-varying linear systems more complicated. Discussions about block pulse function methods for solving analysis and identification problems of time-varying linear systems can be found in the papers of Hwang and Guo (1984a), Jaw and Kung (1984).

7.2.3   Block pulse regression equations

To avoid these disadvantages, the block pulse regression equations of time-varying linear systems can also be derived (Jiang and Schaufelberger, 1991c). Since (7.46) can be rewritten as:

$$
\underbrace{\int_0^t \cdots \int_0^t a_k^{(j)}(t)y(t)\,dt \cdots dt}_{i \text{ times}} =
\begin{pmatrix}
\displaystyle\sum_{s=0}^{n-j}\sum_{r=1}^{1}\frac{(j+s)!h^{i+s}}{(i+s+1)!}a_{k,j+s}y_r p_{i,s,r,1} \\[2.5ex]
\displaystyle\sum_{s=0}^{n-j}\sum_{r=1}^{2}\frac{(j+s)!h^{i+s}}{(i+s+1)!}a_{k,j+s}y_r p_{i,s,r,2} \\[2.5ex]
\vdots \\[1ex]
\displaystyle\sum_{s=0}^{n-j}\sum_{r=1}^{m}\frac{(j+s)!h^{i+s}}{(i+s+1)!}a_{k,j+s}y_r p_{i,s,r,m}
\end{pmatrix}^T
\Phi(t)
$$

$$
= \sum_{l=1}^{m}\sum_{s=0}^{n-j}\sum_{r=1}^{l}\frac{(j+s)!h^{i+s}}{(i+s+1)!}a_{k,j+s}y_r p_{i,s,r,l}\phi_l(t) \tag{7.50}
$$

the left-hand side of (7.43) becomes:

$$
\underbrace{\int_0^t \cdots \int_0^t a_k(t)y^{(k)}(t)\,dt \cdots dt}_{N \text{ times}}
$$

$$
= \sum_{k=0}^{N}\sum_{j=0}^{k}\sum_{l=1}^{m}\sum_{s=0}^{n-j}\sum_{r=1}^{l}(-1)^j\binom{k}{j}\frac{(j+s)!h^{N-k+j+s}}{(N-k+j+s+1)!}a_{k,j+s}y_r p_{N-k+j,s,r,l}\phi_l(t)
$$

$$
+ \sum_{k=0}^{N}\sum_{s=1}^{k}\sum_{j=0}^{k-s}\sum_{l=1}^{m}\sum_{r=1}^{l}(-1)^{j+1}\binom{j+s-1}{s-1}\frac{h^i}{(i+1)!}a_k^{(j)}(0)y^{(k-s-j)}(0)p_{N-s,0,r,l}\phi_l(t) \tag{7.51}
$$

If we substitute $b_k(t)$ for $a_k(t)$ and $u^{(k)}(t)$ for $y^{(k)}(t)$ in the above expression, we also get a similar expression of the right-hand side in (7.43).

After equating each $\phi_l(t)$ $(l = 1, 2, \ldots, m)$ of block pulse series in (7.47) on both sides separately, we obtain m equations of block pulse coefficients. If the lth equation is denoted as $E_{(l)}$, it has the form:

$$
\sum_{k=0}^{N}\sum_{j=0}^{k}\sum_{s=0}^{n-j}\sum_{r=1}^{l}(-1)^j\binom{k}{j}\frac{(j+s)!h^{N-k+j+s}}{(N-k+j+s+1)!}a_{k,j+s}y_r p_{N-k+j,s,r,l}
$$

$$
+ \sum_{k=0}^{N}\sum_{s=1}^{k}\sum_{j=0}^{k-s}\sum_{r=1}^{l}(-1)^{j+1}\binom{j+s-1}{s-1}\frac{h^i}{(i+1)!}a_k^{(j)}(0)y^{(k-s-j)}(0)p_{N-s,0,r,l}
$$

$$
= \sum_{k=0}^{N}\sum_{j=0}^{k}\sum_{s=0}^{n-j}\sum_{r=1}^{l}(-1)^j\binom{k}{j}\frac{(j+s)!h^{N-k+j+s}}{(N-k+j+s+1)!}b_{k,j+s}u_r p_{N-k+j,s,r,l}
$$

$$
+ \sum_{k=0}^{N}\sum_{s=1}^{k}\sum_{j=0}^{k-s}\sum_{r=1}^{l}(-1)^{j+1}\binom{j+s-1}{s-1}\frac{h^i}{(i+1)!}b_k^{(j)}(0)u^{(k-s-j)}(0)p_{N-s,0,r,l} \tag{7.52}
$$

Applying the operation $\displaystyle\sum_{i=0}^{N}(-1)^i\binom{N}{i}E_{(l+N-i)}$ on the $(N+1)$ successive equations $E_{(l)}$,

$E_{(l+1)}, \ldots, E_{(l+N)}$ $(1 \le l \le m - N)$, we obtain:

$$\sum_{i=0}^{N}\sum_{k=0}^{N}\sum_{j=0}^{k}\sum_{s=0}^{n-j}\sum_{r=1}^{l+N-i}(-1)^{i+j}\binom{k}{j}\binom{N}{i}\frac{(j+s)!h^{N-k+j+s}}{(N-k+j+s+1)!}a_{k,j+s}y_r p_{N-k+j,s,r,l+N-i}$$

$$+\sum_{i=0}^{N}\sum_{k=0}^{N}\sum_{s=1}^{k}\sum_{j=0}^{k-s}\sum_{r=1}^{l+N-i}(-1)^{i+j+1}\binom{j+s-1}{s-1}\binom{N}{i}\frac{h^i}{(i+1)!}a_k^{(j)}(0)y^{(k-s-j)}(0)p_{N-s,0,r,l+N-i}$$

$$=\sum_{i=0}^{N}\sum_{k=0}^{N}\sum_{j=0}^{k}\sum_{s=0}^{n-j}\sum_{r=1}^{l+N-i}(-1)^{i+j}\binom{k}{j}\binom{N}{i}\frac{(j+s)!h^{N-k+j+s}}{(N-k+j+s+1)!}b_{k,j+s}u_r p_{N-k+j,s,r,l+N-i}$$

$$+\sum_{i=0}^{N}\sum_{k=0}^{N}\sum_{s=1}^{k}\sum_{j=0}^{k-s}\sum_{r=1}^{l+N-i}(-1)^{i+j+1}\binom{j+s-1}{s-1}\binom{N}{i}\frac{h^i}{(i+1)!}b_k^{(j)}(0)u^{(k-s-j)}(0)p_{N-s,0,r,l+N-i}$$

$$(7.53)$$

Noticing that

$$\sum_{i=0}^{N}\sum_{r=1}^{l+N-i}(-1)^i\binom{N}{i}p_{N-s,0,r,l+N-i}=0 \qquad (7.54)$$

according to (5.110), the equation (7.53) can be reduced to a form without all the initial values:

$$\sum_{k=0}^{N}\sum_{j=0}^{k}\sum_{s=0}^{n-j}\left((-1)^j\binom{k}{j}\frac{(j+s)!h^{N-k+j+s}}{(N-k+j+s+1)!}a_{k,j+s}\right.$$
$$\left[\sum_{i=0}^{N}\sum_{r=1}^{l+N-i}(-1)^i\binom{N}{i}y_r p_{N-k+j,s,r,l+N-i}\right]\Bigg)$$
$$=\sum_{k=0}^{N}\sum_{j=0}^{k}\sum_{s=0}^{n-j}\left((-1)^j\binom{k}{j}\frac{(j+s)!h^{N-k+j+s}}{(N-k+j+s+1)!}b_{k,j+s}\right.$$
$$\left[\sum_{i=0}^{N}\sum_{r=1}^{l+N-i}(-1)^i\binom{N}{i}u_r p_{N-k+j,s,r,l+N-i}\right]\Bigg) \qquad (7.55)$$

Since the sums in both square brackets of the above equation can be simplified further by (5.97), for example:

$$\sum_{i=0}^{N}\sum_{r=1}^{l+N-i}(-1)^i\binom{N}{i}y_r p_{N-k+j,s,r,l+N-i}$$
$$=\sum_{i=0}^{N}\left((-1)^i\binom{N}{i}\sum_{r=1}^{l-1}y_r p_{N-k+j,s,r,l+N-i}\right)+\sum_{i=0}^{N}\left((-1)^i\binom{N}{i}\sum_{r=l}^{l+N-i}y_r p_{N-k+j,s,r,l+N-i}\right)$$
$$=\sum_{r=1}^{l-1}\left(y_r\sum_{i=0}^{N}(-1)^i\binom{N}{i}p_{N-k+j,s,r,l+N-i}\right)+\sum_{i=0}^{N}\left((-1)^i\binom{N}{i}\sum_{r=l}^{l+N-i}y_r p_{N-k+j,s,r,l+N-i}\right)$$
$$=\sum_{i=0}^{N}\left((-1)^i\binom{N}{i}\sum_{r=l}^{l+N-i}y_r p_{N-k+j,s,r,l+N-i}\right) \qquad (7.56)$$

the equation (7.55) becomes:

$$\sum_{i=0}^{N}\sum_{r=l}^{l+N-i}\left((-1)^i\binom{N}{i}y_r\left[\sum_{k=0}^{N}\sum_{j=0}^{k}\sum_{s=0}^{n-j}(-1)^j\binom{k}{j}\right.\right.$$

$$\left. \frac{(j+s)!h^{N-k+j+s}}{(N-k+j+s+1)!}a_{k,j+s}p_{N-k+j,s,r,l+N-i}\right]\right)$$

$$= \sum_{i=0}^{N}\sum_{r=l}^{l+N-i}\left((-1)^{i}\binom{N}{i}u_{r}\left[\sum_{k=0}^{N}\sum_{j=0}^{k}\sum_{s=0}^{n-j}(-1)^{j}\binom{k}{j}\right.\right.$$

$$\left.\left.\frac{(j+s)!h^{N-k+j+s}}{(N-k+j+s+1)!}b_{k,j+s}p_{N-k+j,s,r,l+N-i}\right]\right) \qquad (7.57)$$

Rearranging the sums in both square brackets of the above equation, for example:

$$\sum_{k=0}^{N}\sum_{j=0}^{k}\sum_{s=0}^{n-j}(-1)^{j}\binom{k}{j}\frac{(j+s)!h^{N-k+j+s}}{(N-k+j+s+1)!}a_{k,j+s}p_{N-k+j,s,r,l+N-i}$$

$$= \sum_{k=0}^{N}\sum_{j=0}^{n}\sum_{s=0}^{\min(k,j)}(-1)^{s}\binom{k}{s}\frac{j!h^{N-k+j}}{(N-k+j+1)!}a_{k,j}p_{N-k+s,j-s,r,l+N-i} \qquad (7.58)$$

the equation (7.57) becomes:

$$\sum_{k=0}^{N}\sum_{j=0}^{n}a_{k,j}z_{k,j,l} = \sum_{k=0}^{N}\sum_{j=0}^{n}b_{k,j}v_{k,j,l} \qquad (7.59)$$

where $z_{k,j,l}$ and $v_{k,j,l}$ $(k=0,1,\ldots,N;j=0,1,\ldots,n)$ are linear combinations of the block pulse coefficients y_{l+r} and u_{l+r} $(r=0,1,\ldots,N)$, respectively. The relation between $z_{k,j,l}$ and y_{l+r} has the form:

$$z_{k,j,l} = \sum_{s=0}^{\min(k,j)}\left((-1)^{s}\binom{k}{s}\left[\sum_{i=0}^{N}\sum_{r=0}^{N-i}(-1)^{i}\binom{N}{i}\frac{j!h^{N-k+j}}{(N-k+j+1)!}p_{N-k+s,j-s,l+r,l+N-i}y_{l+r}\right]\right) \qquad (7.60)$$

and the relation between $v_{k,j,l}$ and u_{l+r} has a form similar to (7.60) if we substitute $v_{k,j,l}$ for $z_{k,j,l}$ and u_{l+r} for y_{l+r}. Equation (7.59) shows that a time-varying linear system described by (7.40), (7.41) and (7.42) can also be approximated by a block pulse regression equation. In fact, if all the coefficients $a_{k,j}$ and $b_{k,j}$ $(j=1,2,\ldots,n)$ are zeros, the general forms of (7.59) and (7.60) in the time-varying case are reduced to the special forms of (7.29) and (7.30) in the time-invariant case.

Based on the derived block pulse regression equation, the identification problem can be solved rather straightforwardly. Noticing that the operations in the square brackets of (7.60) are related to the upper triangular part of an $(N+1)$-dimensional submatrix which is constructed by the lth to $(l+N)$th rows and lth to $(l+N)$th columns of the block pulse operational matrix $P_{N-k+s,j-s}$, and noticing that the lower triangular part of this submatrix is zero according to (5.86), the equation (7.60) can be written as:

$$z_{k,j,l} = \sum_{s=0}^{\min(k,j)}\left((-1)^{s}\binom{k}{s}\left[\sum_{i=0}^{N}\sum_{r=0}^{N}(-1)^{i}\binom{N}{i}\frac{j!h^{N-k+j}}{(N-k+j+1)!}p_{N-k+s,j-s,l+r,l+N-i}y_{l+r}\right]\right) \qquad (7.61)$$

or more clearly, written as:

$$z_{k,j,l} = \sum_{s=0}^{\min(k,j)}(-1)^{s}\binom{k}{s}\frac{j!}{(j-s)!}x_{s,l} \qquad (7.62)$$

and

$$x_{s,l} = \sum_{r=0}^{N} y_{l+r} \left[\sum_{i=0}^{N} (-1)^i \binom{N}{i} \frac{(j-s)! h^{N-k+j}}{(N-k+j+1)!} p_{N-k+s,j-s,l+r,l+N-i} \right]$$

$$= \sum_{r=0}^{N} y_{l+r} \left[\sum_{i=0}^{N} (-1)^i \binom{N}{i} p^*_{N-k+s,j-s,l+r,l+N-i} \right] \qquad (7.63)$$

where $p^*_{N-k+s,j-s,l+r,l+N-i}$ is just the entry of block pulse operational matrix $P_{N-k+s,j-s}$ positioned on the $(l+r)$th row and $(l+N-i)$th column. Equation (7.63) shows that the values $x_{s,l}$ ($s = 0,1,\ldots,\min(k,j)$) can be computed from the Nth-order difference of entries in each row of the submatrix and from the block pulse coefficients of $y(t)$, and (7.62) shows that the value $z_{k,j,l}$ is a linear combination of the values $x_{s,l}$ obtained. If we substitute u_{l+r} for y_{l+r} and $v_{k,j,l}$ for $z_{k,j,l}$, we can also get the value $v_{k,j,l}$ in the same way.

In the above discussion, we notice that only the entries of the first $(N+1)$ main diagonals of block pulse operational matrices are involved in the construction of the block pulse regression equation. This feature enables us to simplify expressions. Instead of the complicated expression (5.86), we use the notation $p_{i,j,l,l+\alpha}$ ($\alpha = 0,1,\ldots,N$) for the parts of entries of the same diagonals. As examples, some formulas of $p_{i,j,l,l+\alpha}$ are listed in Table 7.2, which can directly be inserted into (7.63) to obtain $z_{k,j,l}$. After all the terms of $z_{k,j,l}$ and $v_{k,j,l}$ are obtained, the block pulse regression equation (7.59) can be constructed, and the parameters of the original differential equation $a_0(t)$, ..., $a_N(t)$, $b_0(t)$ ..., $b_N(t)$ can be determined. Since the block pulse regression equation contains

	$\alpha = 0$	$\alpha = 1$	$\alpha = 2$	$\alpha = 3$
$p_{0,0,l,l+\alpha}$	1	0	0	0
$p_{1,0,l,l+\alpha}$	1	2	2	2
$p_{2,0,l,l+\alpha}$	1	6	12	18
$p_{3,0,l,l+\alpha}$	1	14	50	110
$p_{0,1,l,l+\alpha}$	$2l-1$	0	0	0
$p_{1,1,l,l+\alpha}$	$3l-2$	$6l-3$	$6l-3$	$6l-3$
$p_{2,1,l,l+\alpha}$	$4l-3$	$24l-14$	$48l-26$	$72l-38$
$p_{3,1,l,l+\alpha}$	$5l-4$	$70l-45$	$250l-145$	$550l-305$
$p_{0,2,l,l+\alpha}$	$3l^2-3l+1$	0	0	0
$p_{1,2,l,l+\alpha}$	$6l^2-8l+3$	$12l^2-12l+4$	$12l^2-12l+4$	$12l^2-12l+4$
$p_{2,2,l,l+\alpha}$	$10l^2-15l+6$	$60l^2-70l+25$	$120l^2-130l+45$	$180l^2-190l+65$
$p_{3,2,l,l+\alpha}$	$15l^2-24l+10$	$210l^2-270l+101$	$750l^2-870l+311$	$1650l^2-1830l+641$

Table 7.2: Entries of diagonals in extended integration operational matrices.

no initial values of the system, this identification procedure can be applied more easily than (7.48).

Example 7.7 Consider a second-order time-varying linear system described by:

$$y^{(2)}(t) + a_{1,0}y^{(1)}(t) + (a_{0,0} + a_{0,1}t + a_{0,2}t^2)y^{(0)}(t) = (b_{0,0} + b_{0,1}t)u^{(0)}(t)$$

with parameters $a_{0,0} = 0.9$, $a_{0,1} = -0.2$, $a_{0,2} = 0.02$, $a_{1,0} = 1.5$, $b_{0,0} = 1.0$ and $b_{0,1} = 0.1$. The simulated output of the system is obtained from the input $u(t) = \sin t + \sin 2t + \sin 3t$. Our problem is to estimate the six unknown parameters from the given input and output data. Here the sampling period is $h = 0.2$.

From (7.62) and (7.63), the terms of $z_{k,j,l}$ and $v_{k,j,l}$ can be determined, and the corresponding block pulse regression equation has the form:

$$
\begin{aligned}
y_{l+2} - 2y_{l+1} + y_l = & -a_{0,0}\frac{h^2}{6}\Big(y_{l+2} + 4y_{l+1} + y_l\Big) \\
& -a_{0,1}\frac{h^3}{24}\Big((4l+5)y_{l+2} + (16l+8)y_{l+1} + (4l-1)y_l\Big) \\
& -a_{0,2}\frac{h^4}{60}\Big((10l^2+25l+16)y_{l+2} + (40l^2+40l+13)y_{l+1} + (10l^2-5l+1)y_l\Big) \\
& -a_{1,0}\frac{h}{2}\Big(y_{l+2} - y_l\Big) + b_{0,0}\frac{h^2}{6}\Big(u_{l+2} + 4u_{l+1} + u_l\Big) \\
& +b_{0,1}\frac{h^3}{24}\Big((4l+5)u_{l+2} + (16l+8)u_{l+1} + (4l-1)u_l\Big)
\end{aligned}
$$

Based on this equation, the estimation results obtained from the recursive least-squares algorithm are listed in Table 7.3.

t	$\hat{a}_{0,0}$	$\hat{a}_{0,1}$	$\hat{a}_{0,2}$	$\hat{a}_{1,0}$	$\hat{b}_{0,0}$	$\hat{b}_{0,1}$
1.00	0.5665	-0.1112	0.1486	1.1673	1.0693	-0.0665
1.20	0.8458	-0.1757	0.0318	1.4462	1.0138	0.0710
1.40	0.8911	-0.1934	0.0200	1.4914	1.0030	0.0949
1.60	0.8982	-0.1977	0.0194	1.4984	1.0010	0.0988
1.80	0.8996	-0.1989	0.0196	1.4998	1.0005	0.0996
2.00	0.8999	-0.1992	0.0197	1.5001	1.0004	0.0998
$\vdots$	$\vdots$	$\vdots$	$\vdots$	$\vdots$	$\vdots$	$\vdots$	$\vdots$
Actual value	0.9	-0.2	0.02	1.5	1.0	0.1

Table 7.3: Recursive estimation results of a time-varying linear system.

7.3   Multi-input multi-output linear systems

A multi-input multi-output (MIMO) time-invariant linear system can be described by a set of differential equations:

$$\sum_{k=0}^{n_i} a_{i,k} y_i^{(k)}(t) = \sum_{j=1}^{M} \sum_{k=0}^{n_i} b_{i,j,k} u_j^{(k)}(t) \tag{7.64}$$

with parameters $a_{i,k}$ and $b_{i,j,k}$. A MIMO time-varying linear system can be described by a set of differential equations:

$$\sum_{k=0}^{N_i} a_{i,k}(t) y_i^{(k)}(t) = \sum_{j=1}^{M} \sum_{k=0}^{N_i} b_{i,j,k}(t) u_j^{(k)}(t) \tag{7.65}$$

where

$$a_{i,k}(t) = a_{i,k,0} + a_{i,k,1} t + \ldots + a_{i,k,n} t^n \tag{7.66}$$

and

$$b_{i,j,k}(t) = b_{i,j,k,0} + b_{i,j,k,1} t + \ldots + b_{i,j,k,n} t^n \tag{7.67}$$

In both system descriptions, $u_j(t)$ ($j = 1, 2, \ldots, M$) is the jth input, $y_i(t)$ ($i = 1, 2, \ldots, N$) is the ith output.

In Sections 7.1 and 7.2, we notice that the same operations are applied separately on the input and output parts of the original differential equation to obtain the corresponding expressions in the block pulse domain. Since each differential equation in the MIMO linear systems (7.64) and (7.65) contains several input parts and one output part which have the same forms, after expanding the input and output signals into their block pulse series respectively:

$$\begin{aligned} u_j(t) &\doteq \begin{pmatrix} u_{j,1} & u_{j,2} & \cdots & u_{j,m} \end{pmatrix} \Phi(t) \\ &= U_j^T \Phi(t) \end{aligned} \tag{7.68}$$

and

$$\begin{aligned} y_i(t) &\doteq \begin{pmatrix} y_{i,1} & y_{i,2} & \cdots & y_{i,m} \end{pmatrix} \Phi(t) \\ &= Y_i^T \Phi(t) \end{aligned} \tag{7.69}$$

the results obtained in the above two sections for SISO cases can be extended in a straightforward way to the MIMO cases.

7.3.1   System analysis

For describing the relation between the block pulse coefficients of input and output signals, linear algebraic equations similar to (7.6), (7.47) and block pulse regression equations similar to (7.22), (7.59) can be obtained for each differential equation of the above MIMO systems. For example, the linear algebraic equation and the block pulse regression equation corresponding to the MIMO time-invariant linear system are:

$$Y_i^T \sum_{k=0}^{n_i} a_{i,n_i-k} P_k - E^T \sum_{k=0}^{n_i-1} \left(y_{i,0}^{(k)} \sum_{q=k}^{n_i-1} a_{i,n_i+k-q} P_q \right)$$

$$\doteq \sum_{j=1}^{M} \left(U_j^T \sum_{k=0}^{n_i} b_{i,j,n_i-k} P_k - E^T \sum_{k=0}^{n_i-1} \left(u_{j,0}^{(k)} \sum_{q=k}^{n_i-1} b_{i,j,n_i+k-q} P_q \right) \right) \quad (7.70)$$

and

$$\sum_{k=0}^{n_i} A_{i,k} y_{i,l+k} = \sum_{j=1}^{M} \sum_{k=0}^{n_i} B_{i,j,k} u_{j,l+k} \quad (7.71)$$

respectively. In (7.70), $y_{i,0}^{(k)} = \left. \dfrac{d^k y_i(t)}{dt^k} \right|_{t=0}$ and $u_{j,0}^{(k)} = \left. \dfrac{d^k u_j(t)}{dt^k} \right|_{t=0}$ are initial values of $y_i(t)$ and $u_j(t)$. In (7.71), $A_{i,k}$ and $B_{i,j,k}$ are determined from the formula (7.23) or from the relations listed in Appendix C, only the subscripts in the expressions should be modified. Based on these equations, block pulse coefficients of the output signals can be determined from those of the input signals.

7.3.2 System identification

Based on the corresponding linear algebraic equations or block pulse regression equations, parameters of both time-invariant and time-varying MIMO linear systems can be estimated. For example, we can set $a_{i,n_i} = 1$ ($i = 1, 2, \ldots, N$) in the differential equation set of the time-invariant linear system (7.64) without loss of generality, and obtain the expressions similar to the linear algebraic equation set (7.12) and the block pulse regression equations (7.29):

$$Y_i^T = -\sum_{k=1}^{n_i} a_{i,n_i-k} Y_i^T P_k + \sum_{j=1}^{M} \sum_{k=0}^{n_i} b_{i,j,n_i-k} U_j^T P_k + \sum_{k=1}^{n_i} c_{i,k} E^T P_{n_i-k} \quad (7.72)$$

and

$$z_{i,n_i,l} = -\sum_{k=0}^{n_i-1} a_{i,k} z_{i,k,l} + \sum_{j=1}^{M} \sum_{k=0}^{n_i} b_{i,j,k} v_{j,k,l} \quad (7.73)$$

In the linear algebraic equation set (7.72), the initial values of the original differential equations (7.64) are involved in the auxiliary parameters:

$$c_{i,k} = \sum_{q=k}^{n_i} \left(a_{i,q} y_{i,0}^{(q-k)} - \sum_{j=1}^{M} b_{i,j,q} u_{j,0}^{(q-k)} \right) \quad (7.74)$$

In the block pulse regression equations (7.73), the terms $v_{j,k,l}$ and $z_{i,k,l}$ are linear combinations of the block pulse coefficients of input and output signals which can also be determined from (7.30) or from the relations listed in Appendix C, only the subscripts in the expressions should be modified. Based on (7.72) or (7.73), system parameters $a_{i,k}$ and $b_{i,j,k}$ can be estimated.

Example 7.8 Consider a two-input one-output linear system described by:

$$G(s) = \left(\frac{b_{1,1,1}s + b_{1,1,0}}{s^2 + a_{1,1}s + a_{1,0}} \quad \frac{b_{1,2,0}}{s^2 + a_{1,1}s + a_{1,0}} \right)$$

with parameters $a_{1,0} = 0.5$, $a_{1,1} = 1.5$, $b_{1,1,0} = 0.5$, $b_{1,1,1} = 1.0$ and $b_{1,2,0} = 1.0$. The simulated output of the system is obtained from the input signals $u_1(t) = \cos t$ and $u_2(t) = \cos 2t$. Our problem is to estimate the unknown parameters from the given input and output data. Here, the sampling period is $h = 0.05$.

According to the relations listed in Appendix C, the equation (7.73) becomes:

$$y_{1,l+2} - 2y_{1,l+1} + y_{1,l} = -\frac{h}{2}(y_{1,l+2} - y_{1,l})a_{1,1} - \frac{h^2}{6}(y_{1,l+2} + 4y_{1,l+1} + y_{1,l})a_{1,0}$$

$$+\frac{h}{2}(u_{1,l+2} - u_{1,l})b_{1,1,1} + \frac{h^2}{6}(u_{1,l+2} + 4u_{1,l+1} + u_{1,l})b_{1,1,0}$$

$$+\frac{h^2}{6}(u_{2,l+2} + 4u_{2,l+1} + u_{2,l})b_{1,2,0}$$

Based on this equation, the estimation results obtained from the recursive least-squares algorithm are listed in Table 7.4. Discussions about block pulse function methods for MIMO linear systems can be found in the papers of Jiang (1987a, 1990).

7.4 Linear systems containing time delays

Containing time delay τ in the input, a time-invariant linear system can be described by the differential equation:

$$\sum_{k=0}^{n} a_k y^{(k)}(t) = \sum_{k=0}^{n} b_k u^{(k)}(t - \tau) \tag{7.75}$$

and a time-varying linear system can be described by the differential equation:

$$\sum_{k=0}^{N} a_k(t)y^{(k)}(t) = \sum_{k=0}^{N} b_k(t)u^{(k)}(t - \tau) \tag{7.76}$$

t	$\hat{a}_{1,0}$	$\hat{a}_{1,1}$	$\hat{b}_{1,1,0}$	$\hat{b}_{1,1,1}$	$\hat{b}_{1,2,0}$
2.5	0.4769	1.5153	0.5009	1.0124	1.0117
3.0	0.4975	1.5021	0.4999	1.0016	1.0019
3.5	0.4999	1.5013	0.5005	1.0007	1.0009
4.0	0.5003	1.5012	0.5006	1.0005	1.0008
True value	0.5	1.5	0.5	1.0	1.0

Table 7.4: Recursive estimation results of a MIMO linear system.

Using an intermediate variable:

$$w^{(k)}(t) = u^{(k)}(t - \tau) \tag{7.77}$$

we have the differential equations:

$$\sum_{k=0}^{n} a_k y^{(k)}(t) = \sum_{k=0}^{n} b_k w^{(k)}(t) \tag{7.78}$$

and

$$\sum_{k=0}^{N} a_k(t) y^{(k)}(t) = \sum_{k=0}^{N} b_k(t) w^{(k)}(t) \tag{7.79}$$

which are equivalent to (7.75), (7.76) respectively and have the same forms as (7.1), (7.40) of the SISO linear systems. Based on them, the block pulse function methods discussed in Sections 7.1 and 7.2 can be extended straightforwardly.

7.4.1 System analysis

Expanding the signals into their block pulse series and setting the width of the block pulses h small enough, it is reasonable to assume that the delay time is qh with a positive integer q. Obviously, the block pulse coefficients of $u(t - \tau)$ are only the shifted block pulse coefficients of $u(t)$:

$$w_k = u_{k-q} \tag{7.80}$$

According to this relation, linear algebraic equations similar to (7.6), (7.47) and block pulse regression equations similar to (7.22), (7.59) can be obtained for the differential equations of the above time delay systems. For example, we can obtain the block pulse difference equation of a time-invariant linear system directly from (7.22):

$$\sum_{k=0}^{n} A_k y_{l+k} = \sum_{k=0}^{n} B_k u_{l+k-q} \tag{7.81}$$

with $l = q + 1, q + 2, \ldots, m - n$, or

$$\sum_{k=0}^{n} A_k y_{l+k+q} = \sum_{k=0}^{n} B_k u_{l+k} \tag{7.82}$$

with $l = 1, 2, \ldots, m - n - q$, from which the block pulse coefficients of the output signal can be evaluated recursively.

7.4.2 System identification

Based on the corresponding linear algebraic equations or block pulse regression equations, parameters of both time-invariant and time-varying linear systems containing time delays can be estimated. For example, after setting $a_n = 1$ in the differential equation of the time-invariant linear system (7.75), we can obtain the expressions similar to the block pulse regression equations (7.29):

$$z_{n+q,l} = -\sum_{k=0}^{n-1} a_k z_{k+q,l} + \sum_{k=0}^{n} b_k v_{k,l} \qquad (7.83)$$

where $z_{k+q,l}$ and $v_{k,l}$ $(k = 0, 1, \ldots, n)$ are linear combinations of the block pulse coefficients y_{l+j+q} and u_{l+j} $(j = 0, 1, \ldots, n)$, respectively. In case the delay time is known, the estimation procedures discussed in Section 7.1 can be applied directly, only the block pulse coefficients of the input part or output part should be shifted by a proper number q. In case the delay time is unknown, it can be estimated together with the parameters by defining certain loss functions which penalize the estimated results and indicate the right delay time. Such ideas have been first used in the identification of discrete-time systems containing time delay, and can be adopted here directly because block pulse regression equations have similar forms as discrete-time systems. After setting a value of delay time, system parameters can be estimated, but these estimation results may have false values because the assumed delay time may be incorrect. The estimations of delay time and system parameters can have their correct values only when the loss functions take their minimum. As a concrete example (Hsia, 1968), if the relation (7.83) is used p times with different block pulse coefficients from the measured data, and if the error of the jth equation $(j = j_1, j_2, \ldots, j_p)$ under the estimated parameters is expressed as:

$$e_j(\tau) = z_{l+n+j+q} + \sum_{k=0}^{n-1} \hat{a}_k z_{l+k+j+q} - \sum_{k=0}^{n} \hat{b}_k v_{l+k+j} \qquad (7.84)$$

the loss function can be defined as:

$$J(\tau) = \frac{1}{p} \sum_{j=1}^{p} \left(e_j(\tau) \right)^2 \qquad (7.85)$$

Based on (7.83), the parameters $\hat{a}_k$ and $\hat{b}_k$ can be estimated. We can repeat this procedure as τ increases each time by h until the loss function $J(\tau)$ in (7.85) reaches its minimum. Under this delay time, we obtain good estimations of the parameters a_k and b_k.

Example 7.9 Consider a second-order time-invariant delay system described by the transfer function:

$$G(s) = \frac{b_0\, e^{-\tau s}}{s^2 + a_1 s + a_0}$$

with parameters $a_0 = 2.0$, $a_1 = 3.0$, $b_0 = 1.0$, and the delay time $\tau = 0.5$. The simulated output of the system is obtained from the input $u(t) = \sin t + \sin 2t$. Our problem is to estimate the unknown parameters from the given input and output data. Here, the sampling period is $h = 0.05$.

To estimate the delay time τ together with the parameters a_0, a_1 and b_0 from the simulated input and output signals, we use the block pulse regression equation (7.32) and the loss function (7.85) with $p = 200$. Under different delay times, the values of the

$\hat{\tau}$	$J(\tau) \times 10^{-12}$	$\hat{a}_0$	$\hat{a}_1$	$\hat{b}_0$
0.40	10201	1.5531	2.2259	0.7692
0.45	2791	1.7466	2.5759	0.8712
0.50	0.05	2.0014	3.0032	1.0010
0.55	3688	2.3446	3.5405	1.1710
0.60	18001	2.8191	4.2364	1.3996
True value		2.0	3.0	1.0

Table 7.5: Estimation results of a time delay linear system.

loss function and the batch processing estimation results of the ordinary least-squares algorithm (7.17) are listed in Table 7.5. Obviously, the values of the loss function indicate that the correct input delay time is $\tau = 0.5$. Discussions about block pulse function methods for linear systems containing time delay can be found in the papers of Wu and Wong (1980), Jiang (1987a, 1990).

7.5   Hammerstein model nonlinear systems

To extend the block pulse function methods to nonlinear systems, Hammerstein model nonlinear system is the simplest case, because a Hammerstein model nonlinear system is constructed by a memoryless nonlinear gain followed by a linear subsystem. Using an intermediate variable $w(t)$, the Hammerstein model nonlinear system can be described by linear differential equations. In the time-invariant case, it is:

$$\sum_{k=0}^{n} a_k y^{(k)}(t) = \sum_{k=0}^{n} b_k w^{(k)}(t) \tag{7.86}$$

and in the time-varying case, it is:

$$\sum_{k=0}^{N} a_k(t) y^{(k)}(t) = \sum_{k=0}^{N} b_k(t) w^{(k)}(t) \tag{7.87}$$

In both cases, the intermediate variable is:

$$w(t) = f(u(t)) \tag{7.88}$$

where $f(*)$ is the nonlinear gain. Since (7.86) and (7.87) have the same forms as (7.1), (7.40) of the SISO linear systems, the results discussed in Sections 7.1 and 7.2 can also be extended straightforwardly.

7.5.1   System analysis

Our problem here is to evaluate the block pulse coefficients of the output signal $y(t)$ in a finite interval $t \in [0, T)$ from the system differential equation (7.86), the nonlinear gain $f(*)$ and the initial values $y_0^{(k)}$ $(k = 0, 1, \ldots, n-1)$ under the input excitation $u(t)$. From the memoryless nonlinear gain, we can obtain a relation between the block pulse coefficients of the input signal $u(t)$ and the intermediate variable $w(t)$:

$$w_k \doteq f(u_k) \tag{7.89}$$

Thus, the analysis problem of Hammerstein model nonlinear systems is the same as SISO linear systems, only the block pulse coefficients of the intermediate variable should be evaluated as an extra step. For example, the block pulse difference equation of a time-invariant Hammerstein model nonlinear system is:

$$\sum_{k=0}^{n} A_k y_{l+k} = \sum_{k=0}^{n} B_k w_{l+k} \tag{7.90}$$

from which the block pulse coefficients of the output signal can be evaluated recursively.

7.5.2   System identification

In most cases, the nonlinear gain $f(*)$ can be approximated by a polynomial with an appropriate order p:

$$f(u(t)) \doteq \sum_{i=1}^{p} r_i \big(u(t)\big)^i \tag{7.91}$$

Under this assumption, the identification problem of Hammerstein model nonlinear systems becomes the estimation of the system parameters and the coefficients in the polynomial of the nonlinear gain from the block pulse coefficients of the input and output signals.

As an example, we use the block pulse regression equation to solve this identification problem in the time-invariant case. After establishing the relation between the block pulse coefficients of the input signal $u(t)$ and the intermediate variable $w(t)$ from (7.91):

$$w_k \doteq \sum_{i=1}^{p} r_i u_k^i \tag{7.92}$$

and defining a set of auxiliary parameters:

$$c_{i,k} = r_i b_k \tag{7.93}$$

where $i = 1, 2, \ldots, p;\ k = 0, 1, \ldots, n$, the block pulse regression equation of the Hammerstein model nonlinear system can be expressed as:

$$z_{n,l} = -\sum_{k=0}^{n-1} a_k z_{k,l} + \sum_{i=1}^{p} \sum_{k=0}^{n} c_{i,k} v_{i,k,l} \tag{7.94}$$

In (7.94), $z_{k,l}$ $(k = 0, 1, \ldots, n)$ are linear combinations of the block pulse coefficients y_{l+j} $(j = 0, 1, \ldots, n)$, but $v_{i,k,l}$ $(k = 0, 1, \ldots, n)$ are linear combinations of the powers of the block pulse coefficients u_{l+j} $(j = 0, 1, \ldots, n)$. The relations between the block pulse coefficients listed in Appendix C can also be applied here, only modifications should be done for obtaining $v_{i,k,l}$ from u_{l+j}, e.g. the modified relations between $v_{i,k,l}$ and u_{l+j} for second-order systems are:

$$v_{i,2,l} = u_{l+2}^i - 2u_{l+1}^i + u_l^i$$

$$v_{i,1,l} = \frac{h}{2}(u_{l+2}^i - u_l^i) \tag{7.95}$$

$$v_{i,0,l} = \frac{h^2}{6}(u_{l+2}^i + 4u_{l+1}^i + u_l^i)$$

Based on (7.94), the parameters a_k $(k = 0, 1, \ldots, n-1)$ and $c_{i,k}$ $(i = 1, 2, \ldots, p; k = 0, 1, \ldots, n)$ can be estimated. Since we can set one of the coefficients in the polynomial (7.91) arbitrarily if it is not zero, e.g. $r_1 = 1$, the system parameters b_k $(k = 0, 1, \ldots, n)$ and the polynomial coefficients r_i $(i = 2, 3, \ldots, p)$ can be determined respectively as:

$$\hat{b}_k = \hat{c}_{1,k} \tag{7.96}$$

and

$$\hat{r}_i = \sum_{k=1}^{n} \hat{b}_k \hat{c}_{i,k} \bigg/ \sum_{k=1}^{n} \hat{b}_k^2 \tag{7.97}$$

Example 7.10 Consider a Hammerstein model nonlinear system with its linear part:

$$y^{(2)}(t) + a_1 y^{(1)}(t) + a_0 y(t) = b_0 w(t)$$

and its nonlinear gain:

$$w(t) = u(t) + r_2 u^2(t) + r_3 u^3(t) + r_4 u^4(t)$$

Here, the parameters are $a_0 = 2.0$, $a_1 = 3.0$ and $b_0 = 1.0$, the coefficients in the power polynomial of the nonlinear gain are $r_2 = 0.5$, $r_3 = -0.08$ and $r_4 = 0.004$. The simulated output of the system is obtained from the input $u(t) = \sin t + \sin 2t$. Our problem is to estimate the unknown parameters from the given input and output data. The sampling period is $h = 0.05$.

Using the relations in Appendix C between $z_{k,l}$ and y_{l+j}, and using the modified relations in (7.95) between $v_{i,k,l}$ and u_{l+j}, we have the block pulse regression equation of this system:

$$y_{l+2} - 2y_{l+1} + y_l = -\frac{h}{2}(y_{l+2} - y_l)a_1 - \frac{h^2}{6}(y_{l+2} + 4y_{l+1} + y_l)a_0$$

$$+\frac{h^2}{6}(u_{l+2} + 4u_{l+1} + u_l)c_{1,0} + \frac{h^2}{6}(u_{l+2}^2 + 4u_{l+1}^2 + u_l^2)c_{2,0}$$

$$+\frac{h^2}{6}(u_{l+2}^3 + 4u_{l+1}^3 + u_l^3)c_{3,0} + \frac{h^2}{6}(u_{l+2}^4 + 4u_{l+1}^4 + u_l^4)c_{4,0}$$

t	$\hat{a}_0$	$\hat{a}_1$	$\hat{b}_0$	$\hat{r}_2$	$\hat{r}_3$	$\hat{r}_4$
3.5000	2.0002	2.9998	0.9990	0.5021	-0.0814	0.0044
4.0000	2.0004	2.9998	0.9996	0.5032	-0.0840	0.0054
4.5000	2.0001	2.9997	0.9996	0.5015	-0.0818	0.0046
5.0000	2.0000	2.9997	0.9991	0.5010	-0.0800	0.0039
True value	2.0	3.0	1.0	0.5	-0.08	0.004

Table 7.6: Recursive estimation results of a Hammerstein model nonlinear system.

From the known coefficient $r_1 = 1$, we also have $b_0 = c_{1,0}$, $r_2 = c_{2,0}/b_0$, $r_3 = c_{3,0}/b_0$ and $r_4 = c_{4,0}/b_0$. Based on these equations, the system parameters together with the coefficients in the power polynomial of the nonlinear gain can be estimated. Using the recursive least-squares algorithm, some estimation results are listed in Table 7.6. Discussions about block pulse function methods for Hammerstein model nonlinear systems can be found in the papers of Jiang (1987b, 1988).

7.6   Inversions of Laplace transforms

Dynamic characteristics of continuous-time systems are usually described by transfer functions, from which many problems about system analysis and design can be solved simply and plainly. But it is not always an easy task to convert the results from the Laplace domain to the time domain, therefore the study about numerical solutions of inverse Laplace transforms is necessary. As one of the methods, we use block pulse functions here to solve this problem in a piecewise constant approximate way.

7.6.1   Functions in general forms

In the block pulse domain, the inverse Laplace transform of functions with general forms, e.g. rational functions, irrational functions and transcendental functions, can be solved by the method of block pulse transform which was discussed in Section 4.1. For a function $F(s)$, this method first expresses $sF(s)$ as $F_1(s^{-1})$ to construct a function $F^s(z)$, and then expands $F^s(z)$ into power series of z^{-1} to obtain the coefficients of its first m terms. Thus we have the block pulse coefficients of the inverse Laplace transform of $F(s)$.

This procedure sounds quite simple, but its practical manipulation is rather complicated. For obtaining the first m coefficients in the power series, the first- to $(m-1)$th-order derivatives of $F^s(z)$ with respect to z^{-1} should be calculated. Since the expression of (4.17) is complicated, it is tedious to calculate the derivatives.

To improve the above procedure, we consider the expression:

$$F(s) = \frac{1}{s} F_1(s^{-1}) \tag{7.98}$$

once more, which means that the inverse Laplace transform of $F(s)$ is a unit step response of the transfer function $F_1(s^{-1})$. Since the block pulse integration operatoral matrix P corresponds to the integrator s^{-1}, and since the block pulse coefficient vector E corresponds to the unit step function, the block pulse series of the inverse Laplace transform of $F(s)$ can be obtained from:

$$\mathcal{L}^{-1}\{F(s)\} \doteq E^T F_1(P)\Phi(t) \tag{7.99}$$

This equation shows that the main work of the improved method is to construct the matrix $F_1(P)$.

Noticing that all the m eigenvalues of the matrix P are $h/2$, the Sylvester interpolation theorem gives:

$$F_1(P) = \sum_{i=0}^{m-1} \frac{d^i F_1(\lambda)}{d\lambda^i} \frac{1}{i!} (P - \lambda I)^i \bigg|_{\lambda = h/2} \tag{7.100}$$

where

$$\left(P - \frac{h}{2}I\right)^i = h^i \left(H + H^2 + \cdots + H^{m-1}\right)^i$$
$$= h^i \left[\binom{i-1}{0}H^i + \binom{i}{1}H^{i+1} + \binom{i+1}{2}H^{i+2} + \cdots + \binom{i+k-1}{k}H^{i+k} + \cdots\right] \tag{7.101}$$

Since the right-hand side of (7.101) contains merely powers of delay matrix, its nonzero entries exist only in a certain upper triangular part and its jth row can be obtained by shifting the first row $(j-1)$ positions to the right. This matrix structure simplifies the construction of (7.100), so that we need only concentrate on those nonzero entries in the first rows of the matrix $F_1(P)$. Denoting γ_i^k as the kth entry of the first row of the matrix in the square brackets of (7.101), we can obtain:

$$\gamma_i^k = \begin{cases} 0 & \text{for } k = 1, 2, \ldots, i \\ \binom{k-2}{i-1} & \text{for } k = i+1, i+2, \ldots, m \end{cases} \tag{7.102}$$

Applying the properties of combinations to (7.102), we can find some relations, which are useful in constructing the matrix $F_1(P)$. As examples, for $k = 2, 3, \ldots, m$, we have:

$$\gamma_1^k = 1 \tag{7.103}$$

for $i = 2, 3, \ldots, k$, we have:

$$\gamma_i^k = \frac{k-i}{i-1} \gamma_{i-1}^k \tag{7.104}$$

and for $k = i, i+1, \ldots, m-1$, we have:

$$\gamma_i^{k+1} = \gamma_i^k + \gamma_{i-1}^k \qquad (7.105)$$

For constructing the matrix $F_1(P)$ from (7.100), the first- to $(m-1)$th-order derivatives of $F_1(\lambda)$ with respect to λ should be calculated. But in comparison with the derivatives in the block pulse transform, the method discussed here is more efficient because the expression $F_1(x)$ is usually simpler than the expression $\dfrac{1}{1-x}F_1\left(\dfrac{h}{2}\dfrac{1+x}{1-x}\right)$.

In fact, these two methods are equivalent (Marszalek, 1985b) for solving the inverse Laplace transform problem based on the block pulse functions.

Example 7.11 Find the block pulse coefficients of the inverse Laplace transform of a transcendental function which appears in a percolation problem:

$$F(s) = \frac{1}{s}e^{-\frac{s}{s+1}}$$

in the interval $t \in [0,8)$.

Set the number of block pulse functions to $m = 4$. Since $F_1(s^{-1}) = e^{-\frac{1}{1+s^{-1}}}$, we first obtain the values of the function and its derivatives:

$$e^{-\frac{1}{1+\lambda}}\Big|_{\lambda=1} = 0.6065$$

$$\frac{d\left(e^{-\frac{1}{1+\lambda}}\right)}{d\lambda}\Bigg|_{\lambda=1} = 0.1516$$

$$\frac{d^2\left(e^{-\frac{1}{1+\lambda}}\right)}{d\lambda^2}\Bigg|_{\lambda=1} = -0.1137$$

$$\frac{d^3\left(e^{-\frac{1}{1+\lambda}}\right)}{d\lambda^3}\Bigg|_{\lambda=1} = 0.1232$$

and then obtain from (7.100) the matrix:

$$F_1(P) = 0.6065 \begin{pmatrix} 1 & 0 & 0 & 0 \\ 0 & 1 & 0 & 0 \\ 0 & 0 & 1 & 0 \\ 0 & 0 & 0 & 1 \end{pmatrix} + 0.3033 \begin{pmatrix} 0 & 1 & 1 & 1 \\ 0 & 0 & 1 & 1 \\ 0 & 0 & 0 & 1 \\ 0 & 0 & 0 & 0 \end{pmatrix}$$

$$- 0.2274 \begin{pmatrix} 0 & 0 & 1 & 1 \\ 0 & 0 & 0 & 1 \\ 0 & 0 & 0 & 0 \\ 0 & 0 & 0 & 0 \end{pmatrix} + 0.1643 \begin{pmatrix} 0 & 0 & 0 & 1 \\ 0 & 0 & 0 & 0 \\ 0 & 0 & 0 & 0 \\ 0 & 0 & 0 & 0 \end{pmatrix}$$

$$= \begin{pmatrix} 0.6065 & 0.3033 & 0.0758 & 0.01264 \\ 0 & 0.6065 & 0.3033 & 0.0758 \\ 0 & 0 & 0.6065 & 0.3033 \\ 0 & 0 & 0 & 0.6065 \end{pmatrix}$$

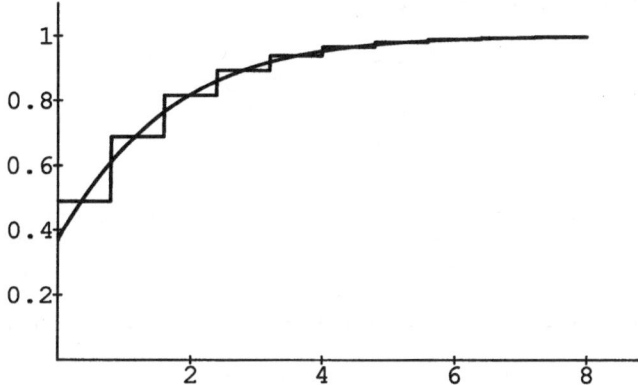

Figure 7.6: Inverse Laplace transform of a transcendental function.

Therefore the block pulse series of this inverse Laplace transform is:

$$\mathcal{L}^{-1}\{\frac{1}{s}e^{-\frac{s}{s+1}}\} \doteq E^T F_1(P)\Phi(t)$$

$$= \left(\begin{array}{cccc} 0.6065 & 0.9098 & 0.9856 & 0.9982 \end{array} \right) \Phi(t)$$

For comparison, the piecewise constant approximate solution with $m = 10$ and its exact solution are illustrated in Figure 7.6.

7.6.2   Functions in special forms

Consider the inverse Laplace transform of a rational transfer function:

$$F(s) = \frac{b_{n-1}s^{n-1} + b_{n-2}s^{n-2} + \cdots + b_0}{s^n + a_{n-1}s^{n-1} + \cdots + a_0} \tag{7.106}$$

In the block pulse domain, the method developed above for general case can also be applied to this special case, but more efficient procedures can be found to avoid the complicated derivative calculations involved. Here, we use the block pulse difference equation to solve this inverse Laplace transform problem.

Since the inverse Laplace transform of $F(s)$ is equivalent to the unit step response of the time-invariant linear system:

$$\frac{Y(s)}{U(s)} = \frac{b_{n-1}s^n + b_{n-2}s^{n-1} + \cdots + b_0 s}{s^n + a_{n-1}s^{n-1} + \cdots + a_0} \tag{7.107}$$

under zero initial values, and since the block pulse coefficients of the unit step input are all ones, we can obtain m equations of block pulse coefficients similar to (7.19):

$$\sum_{i=0}^{n}\left(\frac{h^i}{(i+1)!}a_{n-i}\sum_{j=1}^{l}y_j p_{i,l+1-j}\right) = \sum_{i=0}^{n-1}\left(\frac{h^i}{(i+1)!}b_{n-1-i}\sum_{j=1}^{l}p_{i,l+1-j}\right) \tag{7.108}$$

and the block pulse difference equation similar to (7.22):

$$\sum_{j=0}^{n} A_j y_{l+j} = \sum_{j=0}^{n} B_j \tag{7.109}$$

where A_j is the same as (7.23), but B_j becomes:

$$B_j = \sum_{i=0}^{n-1} \left(\frac{h^i}{(i+1)!} b_{n-1-i} \sum_{k=0}^{n-j} (-1)^k \binom{n}{k} p_{i,n-k-j+1} \right) \tag{7.110}$$

Using these values of B_j, the right-hand side of (7.109) can be expressed as:

$$\sum_{j=0}^{n} B_j = \sum_{i=0}^{n-1} \left(\frac{h^i}{(i+1)!} b_{n-1-i} \left[\sum_{j=0}^{n} \sum_{k=0}^{n-j} (-1)^k \binom{n}{k} p_{i,n-k-j+1} \right] \right) \tag{7.111}$$

According to (5.31), we have the relation for $k = 2, 3, \ldots$:

$$\begin{aligned} \sum_{j=1}^{k} p_{i,j} &= 1 + \sum_{j=2}^{k} j^{i+1} - 2 \sum_{j=1}^{k-1} j^{i+1} + \sum_{j=0}^{k-2} j^{i+1} \\ &= k^{i+1} - (k-1)^{i+1} \end{aligned} \tag{7.112}$$

from which the sum in the square brackets of (7.111) can be rewritten as:

$$\begin{aligned} \sum_{j=0}^{n} \sum_{k=0}^{n-j} (-1)^k \binom{n}{k} p_{i,n-k-j+1} &= \sum_{k=0}^{n} \sum_{j=1}^{n-k+1} (-1)^k \binom{n}{k} p_{i,j} \\ &= (-1)^n \binom{n}{n} p_{i,1} + \sum_{k=0}^{n-1} \left((-1)^k \binom{n}{k} \sum_{j=1}^{n-k+1} p_{i,j} \right) \\ &= \sum_{k=0}^{n} (-1)^k \binom{n}{k} \left((n-k+1)^{i+1} - (n-k)^{i+1} \right) \end{aligned} \tag{7.113}$$

This expression equals zero for $i = 0, 1, \ldots, n-1$ according to (5.34), therefore the block pulse difference equation (7.109) can be simplified as:

$$\sum_{j=0}^{n} A_j y_{l+j} = 0 \tag{7.114}$$

Equation (7.114) indicates that the piecewise constant approximation of the inverse Laplace transform $f(t)$ is determined only by its first n block pulse coefficients f_k ($k = 1, 2, \ldots, n$), which can be evaluated from (7.108):

$$f_k = \frac{\displaystyle\sum_{i=0}^{n-1} \sum_{j=1}^{k} \frac{h^i}{(i+1)!} b_{n-1-i} p_{i,k+1-j} - \sum_{i=0}^{n} \sum_{j=1}^{k-1} \frac{h^i}{(i+1)!} a_{n-i} f_j p_{i,k+1-j}}{\displaystyle\sum_{i=0}^{n} \frac{h^i}{(i+1)!} a_{n-i} p_{i,1}} \tag{7.115}$$

and its remaining block pulse coefficients f_k $(k = n+1, n+2, \ldots, m)$ can be evaluated recursively:

$$f_k = -\frac{\sum_{j=1}^{n} A_{n-j} f_{k-j}}{A_n} \tag{7.116}$$

In each step of evaluation, the size of computation is small because (7.115) is only used in the first n steps with $k = 1, 2, \ldots, n$, and afterwards (7.116) is used recursively.

Example 7.12 Find the block pulse coefficients of the inverse Laplace transform of a rational transfer function:

$$F(s) = \frac{16s^8 + 156s^7 + 648s^6 + 1500s^5 + 2350s^4 + 2857s^3 + 2468s^2 + 1619s + 50}{4s^9 + 28s^8 + 100s^7 + 240s^6 + 405s^5 + 489s^4 + 413s^3 + 215s^2 + 50s}$$

in the interval $t \in [0, 14)$ with $h = 0.3$.

To solve this problem via the block pulse difference equation, the constants $A_0 = -0.4480$, $A_1 = 4.6744$, $A_2 = -21.7238$, $A_3 = 59.0432$, $A_4 = -103.4658$, $A_5 = 121.2726$, $A_6 = -95.1047$, $A_7 = 48.1326$, $A_8 = -14.2686$ and $A_9 = 1.8881$ should be evaluated from the relations in Appendix C. After computing the first nine block pulse coefficients of $f(t)$ from (7.115), the piecewise constant approximation of the result can recursively be computed from (7.116) further. For comparison, this approximate result is illustrated in Figure 7.7 together with the analytical solution:

$$f(t) = 1 + (7 + 2t + 12t^2)e^{-t} - 6e^{-2t} + \big((2 + 5t)\cos(1.5t) - 3t\sin(1.5t)\big)e^{-0.5t}$$

Discussions about block pulse function methods for solving inverse Laplace transform problems can be found in the papers of Chen, Tsay and Wu (1977), Wang (1983), Marszalek (1984b, 1985b), Jiang (1987c).

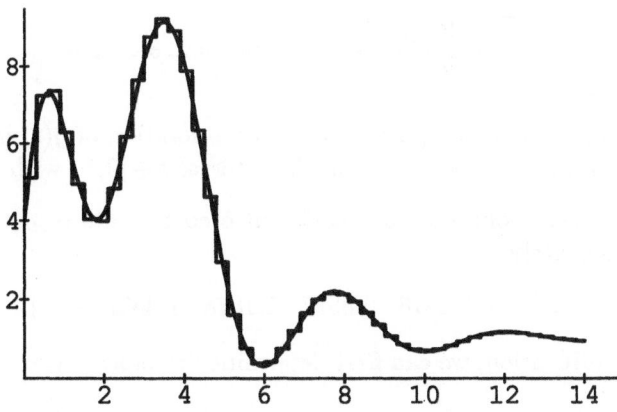

Figure 7.7: Inverse Laplace transform of a rational transfer function.

7.7   Integral equations

As frequently done in the previous sections, problems containing differential equations are first transformed into their equivalent integral forms and then approximated by their corresponding algebraic expressions based on the operation rules of block pulse series. Through such manipulations, the problems containing differential equations can be simplified in the block pulse domain. For problems containing integral equations, similar manipulations can also be done. Here are some of them which can be solved by block pulse function methods.

7.7.1   Deconvolutions

Consider a convolution integral:

$$y(t) = \int_0^t g(\tau)x(t-\tau)d\tau \qquad (7.117)$$

Our problem is to evaluate the block pulse coefficients of $g(t)$ in the interval $t \in [0,T]$ from the known functions $x(t)$ and $y(t)$. Since a process of determining $y(t)$ from $g(t)$ and $x(t)$ is convolution, the process of determining $g(t)$ from $y(t)$ and $x(t)$ is deconvolution.

According to the operation rule of convolution (2.61), the block pulse coefficients of $g(t)$ can be obtained immediately:

$$G^T \doteq Y^T J_X^{-1} \qquad (7.118)$$

From the view of systems science, this deconvolution problem can also be explained as the identification of the impulse response $g(t)$ of an open-loop system from its known input $x(t)$ and output $y(t)$. Based on this consideration, we can first use (B.12) to determine the block pulse transfer matrix from $Y^T = X^T G^{(t)}$, and then evaluate the block pulse coefficients of the impulse response $g(t)$ from the entries of $G^{(t)}$:

$$g_i = \begin{cases} \dfrac{2}{h}g_1^{(t)} & \text{for } i = 1 \\[2ex] \dfrac{2}{h}g_{i-1}^{(t)} - g_{i-1} & \text{for } i = 2,3,\ldots,m \end{cases} \qquad (7.119)$$

Example 7.13 Find the block pulse approximate solution of $g(t)$ in (7.117) from the given $x(t) = e^{-t}$ and $y(t) = t + e^{-t} - 1$ in the interval $t \in [0,5)$ with $m = 10$.

Considered as deconvolution, the block pulse coefficients of $g(t)$ can be obtained from (7.118) immediately:

$$G^T = \begin{pmatrix} 0.1878 & 0.8537 & 1.1878 & 1.8537 & \cdots \end{pmatrix}$$

Considered as identification, we can first determine the block pulse transfer matrix $G^{(t)}$ from (B.12):

$$\begin{pmatrix} g_1^{(t)} & g_2^{(t)} & g_3^{(t)} & g_4^{(t)} & \cdots \end{pmatrix} = \begin{pmatrix} 0.0469 & 0.2604 & 0.5104 & 0.7604 & \cdots \end{pmatrix}$$

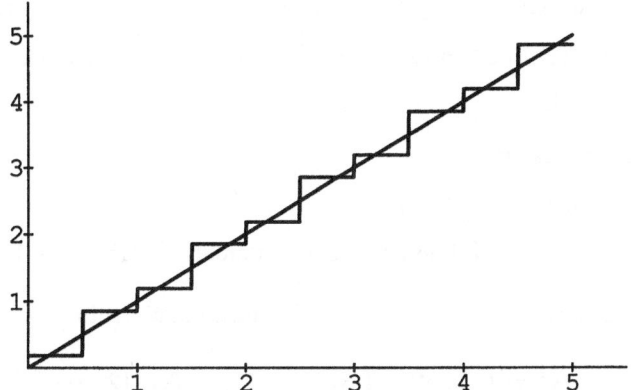

Figure 7.8: Solution of the integral equation (7.117).

and then obtain the same result from (7.119). The piecewise constant approximation together with the analytical solution $g(t) = t$ are illustrated in Figure 7.8.

Consider another convolution integral:

$$y(t) = \int_0^t g(t - \tau) \left(x(\tau) - \int_0^\tau f(\tau - s)y(s)ds \right) d\tau \qquad (7.120)$$

Our problem is to evaluate the block pulse coefficients of $g(t)$ in the interval $t \in [0, T)$ from the known functions $x(t)$, $y(t)$ and $f(t)$.

In the block pulse domain, the manipulation of this problem is similar to the above one. After denoting:

$$r(\tau) = x(\tau) - \int_0^\tau f(\tau - s)y(s)ds \qquad (7.121)$$

we have:

$$R^T \doteq X^T - F^T J_Y \qquad (7.122)$$

Thus the block pulse series of (7.120) is:

$$Y^T \doteq G^T J_R \qquad (7.123)$$

from which the block pulse coefficients of $g(t)$ can be determined.

This deconvolution problem can also be explained as the identification of the impulse response of the feedforward path $g(t)$ in a feedback system from its known input $x(t)$, output $y(t)$ and impulse response of the feedback path $f(t)$. Its block diagram is similar to Figure 6.8 and the relation between signals in the block pulse domain is similar to (6.82). Therefore the block pulse transfer matrix $G^{(t)}$ can first be determined from:

$$Y^T \doteq \left(X^T - Y^T F^{(t)} \right) G^{(t)} \qquad (7.124)$$

and the block pulse coefficients of the impulse response $g(t)$ can then be evaluated from (7.119).

Example 7.14 Find the block pulse approximate solution of $g(t)$ in (7.120) from the given $x(t) = \mu(t)$, $f(t) = e^{-t}$ and $y(t) = \frac{2}{5}\left(1 + (3\sin(t) - \cos(t))e^{-2t}\right)$ in the interval $t \in [0,4)$ with $m = 50$.

From (7.122), we have:

$$R^T \doteq X^T - F^T J_Y$$
$$= \left(\begin{array}{ccccc} 0.9972 & 0.9869 & 0.9688 & 0.9457 & \cdots \end{array}\right)$$

After constructing the matrix J_R, the block pulse coefficients of $g(t)$ can be obtained from (7.123):

$$G^T = \left(\begin{array}{ccccc} 1.8525 & 1.3137 & 1.1751 & 0.7812 & \cdots \end{array}\right)$$

The piecewise constant approximation together with the analytical solution $g(t) = 2e^{-3t}$ are illustrated in Figure 7.9 for comparison. Although small oscillatory behaviour exists in this solution, the average phenomenon of piecewise segments is quite close to the exact curve.

For this deconvolution problem, better result without small oscillations can also be obtained from block pulse functions. Noticing that this problem is solved by (7.122) and (7.123) in two steps, and approximations are introduced in these steps separately, the improvement can be achieved through transforming multiple convolution integrals into algebraic operations in one step. Instead of using the earlier defined convolution operational matrices (2.62) twice, we can derive a more closer block pulse operational matrix for $\phi_i(t) * \phi_j(t) * \phi_k(t)$ $(i,j,k = 1,2,\ldots,m)$ based on the exact calculation:

$$\phi_i(t) * \phi_j(t) * \phi_k(t)$$
$$= \frac{1}{2}\left(t - (i+j+k-3)h\right)^2 \mu(t - (i+j+k-3)h)$$
$$- \frac{3}{2}\left(t - (i+j+k-2)h\right)^2 \mu(t - (i+j+k-2)h)$$

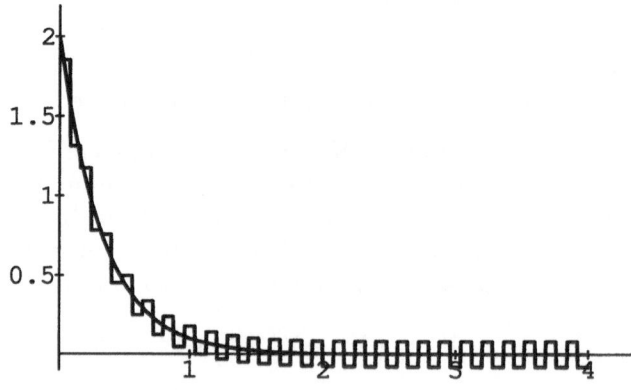

Figure 7.9: Solution of the integral equation (7.120).

$$+ \frac{3}{2}\left(t - (i+j+k-1)h\right)^2 \mu(t - (i+j+k-1)h)$$

$$- \frac{1}{2}\left(t - (i+j+k)h\right)^2 \mu(t - (i+j+k)h) \tag{7.125}$$

The details about this operational matrix can be found in the paper of Hwang and Guo (1984b).

7.7.2 Fredholm integral equations

Consider a Fredholm integral equation:

$$a(x)y(x) = f(x) + \lambda \int_\alpha^\beta k(x,t)y(t)dt \tag{7.126}$$

Our problem is to determine the block pulse coefficients of $y(x)$ in the interval $x \in [\alpha, \beta)$ from the known functions $a(x)$, $f(x)$ and the kernel $k(x,t)$. Usually, we set $\alpha = 0$ to facilitate the use of block pulse functions. In case of $\alpha \neq 0$, we can use (1.3) to change the scale of variable.

Intentionally, we set the numbers of both block pulse functions with respect to x and t as m in this integral equation problem. Since $Y^T\Phi(x)$, $A^T\Phi(x)$, $F^T\Phi(x)$ are respectively block pulse series of the functions $y(x)$, $a(x)$, $f(x)$, and $\Phi^T(x)K\Psi(t)$ is the two-dimensional block pulse series of the function $k(x,t)$, the block pulse series of (7.126) has the form:

$$Y^T D_A \Phi(x) = F^T\Phi(x) + \lambda \int_0^{mh} \Phi^T(x)K\Psi(t)\Psi^T(t)Y\,dt$$

$$= F^T\Phi(x) + \lambda Y^T \int_0^{mh} \Psi(t)\Psi^T(t)dt K^T\Phi(x) \tag{7.127}$$

Noticing that (2.25) gives:

$$\int_0^{mh} \Psi(t)\Psi^T(t)dt = hI \tag{7.128}$$

the above equation becomes:

$$Y^T D_A \Phi(x) = F^T\Phi(x) + \lambda h Y^T K^T\Phi(x) \tag{7.129}$$

from which the block pulse coefficients of $y(x)$ can be determined:

$$Y^T = F^T \left(D_A - \lambda h K^T\right)^{-1} \tag{7.130}$$

Example 7.15 Find the block pulse approximate solution of $y(x)$ from the Fredholm integral equation (7.126) in the interval $x \in [0,1)$ with $m = 10$. The given functions are $a(x) = 1$, $k(x,t) = t - x$, $\lambda = 1$ and $f(x) = 3x/2 - 1/3$.

Since we have the vectors and matrices:

$$D_A = I$$

$$F^T = \left(\begin{array}{ccccc} -0.2583 & -0.1083 & 0.0417 & 0.1917 & \cdots \end{array} \right)$$

and

$$K = \begin{pmatrix} 0 & 0.1000 & 0.2000 & 0.3000 & \cdots \\ -0.1000 & 0 & 0.1000 & 0.2000 & \cdots \\ -0.2000 & -0.1000 & 0 & 0.1000 & \cdots \\ -0.3000 & -0.2000 & -0.1000 & 0 & \cdots \\ \vdots & \vdots & \vdots & \vdots & \ddots \end{pmatrix}$$

the equation (7.130) gives:

$$Y^T = \left(\begin{array}{ccccc} 0.0489 & 0.1490 & 0.2490 & 0.3491 & \cdots \end{array} \right)$$

This piecewise constant approximation together with the analytical solution $y(x) = x$ are illustrated in Figure 7.10 for comparison.

7.7.3 Volterra integral equations

Consider a Volterra integral equation:

$$a(x)y(x) = f(x) + \lambda \int_\alpha^x k(x,t)y(t)dt \tag{7.131}$$

Our problem is to determine the block pulse coefficients of $y(x)$ in the interval $x \in [\alpha, \beta]$ from the known functions $a(x)$, $f(x)$ and the kernel $k(x,t)$. Like in the discussion about the Fredholm integral equation, we also set $\alpha = 0$ here.

Similar to (7.127), the block pulse series of the Volterra integral equation has the form:

$$Y^T D_A \Phi(x) = F^T \Phi(x) + \lambda Y^T \int_0^x \Psi(t)\Psi^T(t)dt K^T \Phi(x) \tag{7.132}$$

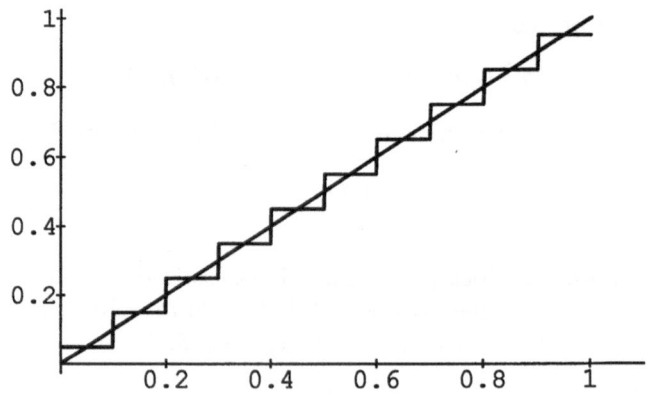

Figure 7.10: Solution of the integral equation (7.126).

After denoting K_i for the ith row of the constant matrix K^T and R_j for the jth row of the conventional interation operational matrix P, the relations (2.25), (2.26) and (2.33) give:

$$\int_0^x \Psi(t)\Psi^T(t)dt K^T \Phi(x)$$

$$= \begin{pmatrix} R_1\Phi(x) & 0 & \cdots & 0 \\ 0 & R_2\Phi(x) & \cdots & 0 \\ \vdots & \vdots & \ddots & \vdots \\ 0 & 0 & \cdots & R_m\Phi(x) \end{pmatrix} \begin{pmatrix} K_1 \\ K_2 \\ \vdots \\ K_m \end{pmatrix} \Phi(x)$$

$$= \begin{pmatrix} R_1\Phi(x)K_1\Phi(x) \\ R_2\Phi(x)K_2\Phi(x) \\ \vdots \\ R_m\Phi(x)K_m\Phi(x) \end{pmatrix}$$

$$= \begin{pmatrix} R_1\Phi(x)\Phi^T(x)K_1^T \\ R_2\Phi(x)\Phi^T(x)K_2^T \\ \vdots \\ R_m\Phi(x)\Phi^T(x)K_m^T \end{pmatrix}$$

$$= \begin{pmatrix} R_1 D_{K_1} \\ R_2 D_{K_2} \\ \vdots \\ R_m D_{K_m} \end{pmatrix} \Phi(x) \tag{7.133}$$

so that (7.132) can be simplified as:

$$Y^T D_A \Phi(x) = F^T \Phi(x) + \lambda Y^T \begin{pmatrix} R_1 D_{K_1} \\ R_2 D_{K_2} \\ \vdots \\ R_m D_{K_m} \end{pmatrix} \Phi(x) \tag{7.134}$$

From this equation we can determine the block pulse coefficients of $y(x)$:

$$Y^T = F^T \left(D_A - \lambda \begin{pmatrix} R_1 D_{K_1} \\ R_2 D_{K_2} \\ \vdots \\ R_m D_{K_m} \end{pmatrix} \right)^{-1} \tag{7.135}$$

Example 7.16 Find the block pulse approximate solution of $y(x)$ from the Volterra integral equation (7.131) in the interval $x \in [0,3)$ with $m = 10$. The given functions are $a(x) = 1$, $k(x,t) = \sinh(t - x)$, $\lambda = 1$ and $f(x) = 3x$.

Since we have the vectors and matrices:

$$D_A = I$$

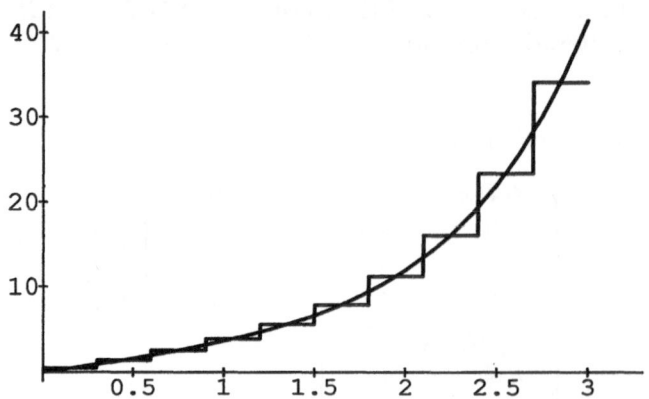

Figure 7.11: Solution of the integral equation (7.131).

$$F^T = \left(\begin{array}{cccc} 0.4500 & 1.3500 & 2.2500 & 3.1500 & \cdots \end{array} \right)$$

and

$$K = \left(\begin{array}{ccccc} 0 & -0.3068 & -0.6414 & -1.0342 & \cdots \\ 0.3068 & 0 & -0.3068 & -0.6414 & \cdots \\ 0.6414 & 0.3068 & 0 & -0.3068 & \cdots \\ 1.0342 & 0.6414 & 0.3068 & 0 & \cdots \\ \vdots & \vdots & \vdots & \vdots & \ddots \end{array} \right)$$

the equation (7.135) gives:

$$Y^T = \left(\begin{array}{cccc} 0.4500 & 1.3914 & 2.4647 & 3.7842 & \cdots \end{array} \right)$$

This piecewise constant approximation together with the analytical solution:

$$y(x) = \frac{3}{2}x + \frac{3\sqrt{2}}{4}\sinh(\sqrt{x})$$

are illustrated in Figure 7.11 for comparison. Discussions about block pulse function methods for solving integral equations can be found in the papers of Kung and Chen (1978), Kwong and Chen (1981b), Wang and Shih (1982), Hwang and Guo (1984b).

Chapter 8

State space representations of dynamic systems

In the previous chapter, we discussed the block pulse function methods for solving problems of continuous-time systems under the input-output representation. In those cases, differential equations of high orders were transformed approximately into their corresponding algebraic forms based on the operation rules of block pulse series, so that the numerical solutions of problems can be obtained more directly. The same ideas can also be applied to solve problems of continuous-time systems which are described by state equations, because the state space representions of such systems are sets of first-order differential equations. But in contrast to the expressions in the previous chapter, the forms of algebraic equations are changed in the block pulse function methods. Instead of scalar parameters, parameter matrices must now be manipulated. And instead of multiple integrals, only one time integrals are involved in this chapter.

8.1 Time-invariant linear systems

As a basic case, we consider first a time-invariant linear system which is described by the state equation:

$$\dot{X}(t) = AX(t) + BU(t) \tag{8.1}$$

and the output equation:

$$Y(t) = CX(t) + DU(t) \tag{8.2}$$

where $U(t)$, $X(t)$ and $Y(t)$ are r-, n- and p-dimensional vectors of input, state and output, and A, B, C and D are constant matrices of appropriate dimensions, respectively.

8.1.1 System analysis

In the block pulse domain, the system analysis problem of a continuous time-invariant linear system under state space representation is similar to that of the input-output representation, i.e. the block pulse coefficients of the output vector $Y(t)$ should be evaluated from the system equations (8.1), (8.2) and the initial value of the states $X(0)$ under the input excitations $U(t)$ in a finite interval $t \in [0, T)$. Usually, we consider only the problem of the state equation (8.1), because after the block pulse coefficients of the state vector are obtained, the block pulse coefficients of the output can easily be determined from the output equation (8.2).

For this purpose, we integrate first the state equation (8.1) from 0 to t on both sides:

$$X(t) - X(0) = A \int_0^t X(t)dt + B \int_0^t U(t)dt \tag{8.3}$$

After expanding the input vector $U(t)$ and the state vector $X(t)$ into their block pulse series respectively:

$$U(t) \doteq \left(\begin{array}{cccc} U_1 & U_2 & \cdots & U_m \end{array} \right) \Phi(t) \tag{8.4}$$

and

$$X(t) \doteq \left(\begin{array}{cccc} X_1 & X_2 & \cdots & X_m \end{array} \right) \Phi(t) \tag{8.5}$$

we obtain the equation:

$$\left(\begin{array}{cccc} X_1 & X_2 & \cdots & X_m \end{array} \right) \Phi(t) - \left(\begin{array}{cccc} X(0) & X(0) & \cdots & X(0) \end{array} \right) \Phi(t)$$
$$\doteq A \left(\begin{array}{cccc} X_1 & X_2 & \cdots & X_m \end{array} \right) P\Phi(t) + B \left(\begin{array}{cccc} U_1 & U_2 & \cdots & U_m \end{array} \right) P\Phi(t) \tag{8.6}$$

Based on the special upper triangular form of the conventional integration operational matrix P, we obtain m equations of block pulse coefficients by equating the block pulse coefficients of $\phi_k(t)$ $(k = 1, 2, \ldots, m)$ on both sides of the above equation separately, e.g. the kth equation $E_{(k)}$:

$$X_k - X(0) = \frac{h}{2} (AX_k + BU_k) + h \sum_{j=1}^{k-1} (AX_j + BU_j) \tag{8.7}$$

from which the block pulse coefficients of the state vector can be evaluated successively:

$$X_k = \Gamma \left(X(0) + \frac{h}{2} BU_k + h \sum_{j=1}^{k-1} (AX_j + BU_j) \right) \tag{8.8}$$

where

$$\Gamma = \left(I - \frac{h}{2} A \right)^{-1} \tag{8.9}$$

But as a disadvantage of this formula for solving the system analysis problem, large size computations will be involved, especially when the number of block pulse functions m is large. This is because the evaluation of each term X_k always begins from the first term X_1. The higher the position of the term X_k, the larger the computation that must be done. Similar to the discussion in Section 7.1, block pulse difference equations can also be introduced here to avoid this disadvantage.

Applying the operation $E_{(k+1)} - E_{(k)}$ $(k = 1, 2, \ldots, m-1)$ on the two neighbouring equations, we obtain:

$$X_{k+1} - X_k = \frac{h}{2} A (X_{k+1} + X_k) + \frac{h}{2} B (U_{k+1} + U_k) \tag{8.10}$$

Based on this, the block pulse coefficients of the state vector $X(t)$ can be computed recursively with $k = 1, 2, \ldots, m - 1$:

$$X_{k+1} = \Gamma \left(\left(I + \frac{h}{2}A \right) X_k + \frac{h}{2} B \left(U_{k+1} + U_k \right) \right) \qquad (8.11)$$

This equation expresses the relation between the input vector and the state vector of a time-invariant linear system in a piecewise constant approximate way, therefore it can be regarded as the block pulse difference equation of (8.1). To start the recursion, X_1 should first be evaluated from (8.8) as the initial value of the block pulse difference equation:

$$X_1 = \Gamma \left(X(0) + \frac{h}{2} BU_1 \right) \qquad (8.12)$$

In the whole recursion, only one n-dimensional matrix must be inverted because Γ is a constant matrix.

Example 8.1 Determine the block pulse coefficients of the state variables of a time-invariant linear system described by (8.1) in the interval $t \in [0, 1)$ with $m = 10$. Here, the parameter matrices are:

$$A = \begin{pmatrix} 1 & 2 \\ 3 & -4 \end{pmatrix}, \qquad B = \begin{pmatrix} 2 & 0 \\ 1 & 1 \end{pmatrix}$$

the initial values of the states are $x_1(0) = 1$, $x_2(0) = 1$, and both input excitations are unit step functions.

Using the relations (8.11) and (8.12), the state vector $X(t)$ can be approximated by the block pulse series:

$$\begin{aligned}
X(t) &\doteq \begin{pmatrix} 1.2711 \\ 1.0756 \end{pmatrix} \phi_1(t) + \begin{pmatrix} 1.8629 \\ 1.2755 \end{pmatrix} \phi_2(t) \\
&+ \begin{pmatrix} 2.5692 \\ 1.5710 \end{pmatrix} \phi_3(t) + \begin{pmatrix} 3.4221 \\ 1.9629 \end{pmatrix} \phi_4(t) + \cdots
\end{aligned}$$

The analytical solutions of the state variables are:

$$x_1(t) = \frac{16}{7}e^{2t} - \frac{3}{35}e^{-5t} - \frac{6}{5}$$

and

$$x_2(t) = \frac{8}{7}e^{2t} + \frac{9}{35}e^{-5t} - \frac{2}{5}$$

which are illustrated in Figure 8.1 together with their piecewise constant approximations for comparison.

Based on the block pulse difference equation discussed above, the transition matrix of the state equation (8.1) can easily be expressed by its block pulse series approximation. After setting $B = 0$, we have the state vector:

$$X(t) = e^{At} X(0) \qquad (8.13)$$

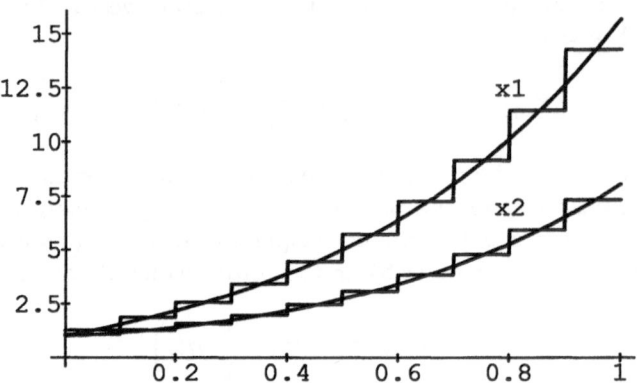

Figure 8.1: State variables of a time-invariant linear system.

Since (8.11) and (8.12) give:

$$X_k = \Lambda^{k-1}\Gamma X(0) \tag{8.14}$$

where

$$\Lambda = \Gamma\left(I + \frac{h}{2}A\right) \tag{8.15}$$

the block pulse series approximation of the system transition matrix is:

$$e^{At} \doteq \sum_{k=1}^{m} \Lambda^{k-1}\Gamma\phi_k(t) \tag{8.16}$$

Equation (8.14) shows that the block pulse coefficients of the transition matrix can be computed successively. In computing all these block pulse coefficients, only one n-dimensional matrix must be inverted.

Example 8.2 Determine the block pulse coefficients of the system transition matrix of Example 8.1. The approximation interval is $t \in [0, 1)$ and the number of block pulse function is $m = 20$.

Since

$$\Gamma = \left(I - \frac{h}{2}A\right)^{-1} = \begin{pmatrix} 1.0292 & 0.0468 \\ 0.0702 & 0.9123 \end{pmatrix}$$

and

$$\Lambda = \Gamma\left(I + \frac{h}{2}A\right) = \begin{pmatrix} 1.0585 & 0.0936 \\ 0.1404 & 0.8246 \end{pmatrix}$$

the result can be obtained from the successive computations $\alpha_1 = \Gamma$, $\alpha_2 = \Lambda\alpha_1$, ..., $\alpha_m = \Lambda\alpha_{m-1}$:

$$e^{At} \doteq \begin{pmatrix} 1.0292 & 0.0468 \\ 0.0702 & 0.9123 \end{pmatrix}\phi_1(t) + \begin{pmatrix} 1.0960 & 0.1349 \\ 0.2023 & 0.7588 \end{pmatrix}\phi_2(t)$$
$$+ \begin{pmatrix} 1.1790 & 0.2138 \\ 0.3206 & 0.6446 \end{pmatrix}\phi_3(t) + \begin{pmatrix} 1.2780 & 0.2866 \\ 0.4299 & 0.5615 \end{pmatrix}\phi_4(t) + \cdots$$

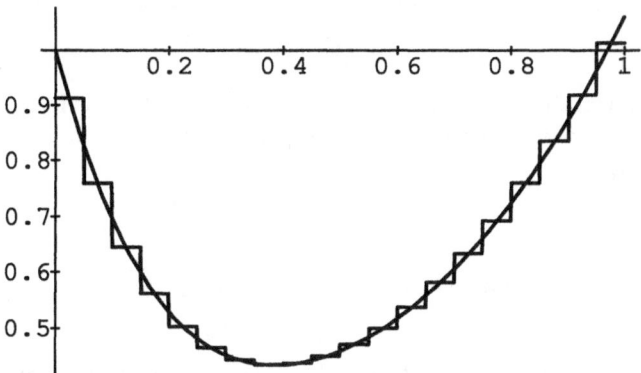

Figure 8.2: Entry in transition matrix of a time-invariant linear system.

In Figure 8.2, the block pulse series approximation of the entry in the second row and second column of the transition matrix is illustrated together with its analytical solution $1/7e^{2t} + 6/7e^{-5t}$.

8.1.2   System identification

In the block pulse domain, the identification problem of a continuous time-invariant linear system is to estimate the system parameter matrices A and B from the block pulse coefficients of the inputs $U(t)$ and the states $X(t)$. This statement implies that all the state variables are measurable in our identification problem.

The unknown parameter matrices can be determined directly from an algebraic equation set which is constructed from the block pulse coefficients of the input and state vectors in (8.6), but similar to the discussion of Section 7.1, such manipulation will enlarge the size of computation and complicate the estimation procedures due to the existence of the operational matrix P and the initial value of the states $X(0)$. To avoid these disadvantages, we can also use the derived relation between the inputs and the states in the block pulse regression equation (8.10) to estimate the unknown parameter matrices. For example, if we denote the matrices:

$$\Theta^T = \begin{pmatrix} A & B \end{pmatrix} \tag{8.17}$$

$$G^T = \frac{h}{2} \begin{pmatrix} X_2 + X_1 & X_3 + X_2 & \cdots & X_m + X_{m-1} \\ U_2 + U_1 & U_3 + U_2 & \cdots & U_m + U_{m-1} \end{pmatrix} \tag{8.18}$$

and

$$F^T = \begin{pmatrix} X_2 - X_1 & X_3 - X_2 & \cdots & X_m - X_{m-1} \end{pmatrix} \tag{8.19}$$

all the $(m-1)$ equations in (8.10) can be expressed in a matrix form:

$$F = G\Theta \tag{8.20}$$

In case of $m - 1 \geq n + r$, the ordinary least-squares algorithm gives:

$$\hat{\Theta} = \left(G^T G\right)^{-1} G^T F \tag{8.21}$$

where $\hat{\Theta}$ is the estimated value of the parameter matrix Θ. This least-squares estimation can also be carried out recursively. After denoting the kth row of the matrix G as the column vector g_k:

$$g_k = \frac{h}{2}\left(\begin{array}{c} X_{k+1} + X_k \\ U_{k+1} + U_k \end{array}\right) \tag{8.22}$$

and the kth row of the matrix F as the row vector f_k:

$$f_k^T = X_{k+1} - X_k \tag{8.23}$$

the parameter matrix Θ can be estimated by the recursive least-squares algorithm:

$$\begin{cases} \hat{\Theta}_k = \hat{\Theta}_{k-1} + \gamma_k S_{k-1} g_k \left(f_k - g_k^T \hat{\Theta}_{k-1}\right) \\ S_k = \dfrac{1}{\alpha}\left(S_{k-1} - \gamma_k S_{k-1} g_k g_k^T S_{k-1}\right) \\ \gamma_k = \dfrac{1}{\alpha + g_k^T S_{k-1} g_k} \end{cases} \tag{8.24}$$

where $\hat{\Theta}_{k-1}$ and $\hat{\Theta}_k$ are the $(k-1)$th and kth estimated values of Θ respectively and α is the forgetting factor, $0 < \alpha \leq 1$.

Example 8.3 Consider a linear system described by (8.1) with parameter matrices:

$$A = \left(\begin{array}{cc} 0 & 1 \\ -0.5 & -1 \end{array}\right), \quad B = \left(\begin{array}{cc} 2 & 1 \\ 1 & 1 \end{array}\right),$$

The inputs are $u_1(t) = \sin t$ and $u_2(t) = \sin 2t$. Our problem is to estimate the unknown parameter matrices from the simulated data of inputs and states. Here the sampling period is $h = 0.01$.

Using the relation between the inputs and the states in (8.10), estimation results can be obtained, e.g. the estimations from the recursive least-squares algorithm (8.24) are listed in Table 8.1 which are quite satisfactory. Discussions about block pulse function methods for solving analysis, identification and state estimation problems of time-invariant linear systems can be found in the papers of Sannuti (1977), Shieh, Yates and Navarro (1978), Shieh, Yeung and McInnis (1978), Dalton (1978), Shieh and Yates (1979), Wu and Juang (1980), Kawaji (1983), Tzafestas, Papastergiou and Anoussis (1984), Sinha and Zhou (1984), Zhang and Chen (1990).

8.1.3 Optimal control

Consider an optimal control problem for the time-invariant linear system (8.1) with respect to a quadratic performance index over a finite interval of time:

$$J = \frac{1}{2}\int_0^T \left(X^T(t)QX(t) + U^T(t)RU(t)\right)dt \tag{8.25}$$

t	0.10	0.15	0.20	0.25	0.30	True value
$\hat{a}_{11}$	-0.0293	-0.0109	-0.0011	0.0000	0.0001	0.0
$\hat{a}_{12}$	1.0220	1.0081	1.0008	1.0000	0.9999	1.0
$\hat{a}_{21}$	-0.5068	-0.5018	-0.4998	-0.4998	-0.4999	-0.5
$\hat{a}_{22}$	-0.9949	-0.9986	-1.0001	-1.0001	-1.0001	-1.0
$\hat{b}_{11}$	1.1875	1.8478	1.9907	1.9997	2.0000	2.0
$\hat{b}_{12}$	1.5237	1.1193	1.0089	1.0000	0.9996	1.0
$\hat{b}_{21}$	0.7807	0.9700	0.9995	0.9993	0.9996	1.0
$\hat{b}_{22}$	1.1370	1.0223	0.9996	0.9999	0.9997	1.0

Table 8.1: Recursive estimation results of a time-invariant linear system.

where Q is a positive semi-definite n-dimensional constant matrix and R is a positive definite r-dimensional constant matrix. For this problem, the optimal feedback control law is (Bryson and Ho, 1975):

$$U(t) = R^{-1}B^T G(t) \tag{8.26}$$

where the n-dimensional adjoint variable $G(t)$ satisfies the canonical equation:

$$\begin{pmatrix} \dot{X}(t) \\ \dot{G}(t) \end{pmatrix} = F \begin{pmatrix} X(t) \\ G(t) \end{pmatrix} \tag{8.27}$$

under the two-point boundary values $X(0) = X_0$ and $G(T) = 0$. In this canonical equation, The coefficient matrix F is:

$$F = \begin{pmatrix} A & BR^{-1}B^T \\ Q & -A^T \end{pmatrix} \tag{8.28}$$

To avoid this two-point boundary value problem in solving $G(t)$, we set the $2n$-dimensional transition matrix of (8.27) as:

$$\Psi(T,t) = \begin{pmatrix} \Psi_{11}(T,t) & \Psi_{12}(T,t) \\ \Psi_{21}(T,t) & \Psi_{22}(T,t) \end{pmatrix} \tag{8.29}$$

where all the submatrices $\Psi_{11}(T,t)$, $\Psi_{12}(T,t)$, $\Psi_{21}(T,t)$ and $\Psi_{22}(T,t)$ are n-dimensional. Noticing that

$$\Psi(T,t) \begin{pmatrix} X(t) \\ G(t) \end{pmatrix} = \begin{pmatrix} X(T) \\ G(T) \end{pmatrix} \tag{8.30}$$

and $G(T) = 0$, we have:

$$G(t) = -\Psi_{22}^{-1}(T,t)\Psi_{21}(T,t)X(t) \tag{8.31}$$

Thus, the optimal feedback control law (8.26) can be expressed as:

$$U(t) = -K(t)X(t) \tag{8.32}$$

with the gain matrix:

$$K(t) = R^{-1}B^T\Psi_{22}^{-1}(T,t)\Psi_{21}(T,t) \tag{8.33}$$

After differentiating (8.30) with respect to t on both sides:

$$\Psi(T,t)\begin{pmatrix} \dot{X}(t) \\ \dot{G}(t) \end{pmatrix} + \dot{\Psi}(T,t)\begin{pmatrix} X(t) \\ G(t) \end{pmatrix} = 0 \tag{8.34}$$

and using (8.27), we have:

$$\dot{\Psi}(T,t)\begin{pmatrix} X(t) \\ G(t) \end{pmatrix} = -\Psi(T,t)F\begin{pmatrix} X(t) \\ G(t) \end{pmatrix} \tag{8.35}$$

Since (8.35) is true in the whole interval $t \in [0,T]$, we obtain the relation:

$$\dot{\Psi}(T,t) = -\Psi(T,t)F \tag{8.36}$$

from which the submatrices $\Psi_{21}(T,t)$ and $\Psi_{22}(T,t)$ can be determined with $\Psi(T,T) = I$.

In applying block pulse functions in this problem, a suboptimal solution with piecewise constant feedback gains can be obtained:

$$K(t) \doteq \sum_{i=1}^{m} K_i\phi_i(t) \tag{8.37}$$

which will converge to the theoretical optimal solution of the same problem when the number of block pulse functions approaches infinity (Wang, 1990). Noticing that the function $\Psi(T,t)$ is related to the variable $(T-t)$, it can be expanded into the block pulse series:

$$\begin{aligned} \Psi(T,t) &\doteq \sum_{i=1}^{m} \Psi_i\phi_i(T-t) \\ &= \sum_{i=1}^{m} \Psi_{m-i+1}\phi_i(t) \end{aligned} \tag{8.38}$$

Since the block pulse coefficients of the transition matrix Ψ_i $(i = 1, 2, \ldots, m)$ can be computed iteratively from (8.16), the piecewise constant approximation of the time-varying gain matrix can be determined as:

$$K_i = R^{-1}B^T\Psi_{22,m-i}^{-1}\Psi_{21,m-i} \tag{8.39}$$

Example 8.4 Determine the piecewise constant feedback gains of the above optimal control problem. In the state equation (8.1), the matrices are:

$$A = \begin{pmatrix} 0 & 0 \\ 1 & 0 \end{pmatrix}, \quad B = \begin{pmatrix} 1 \\ 0 \end{pmatrix}$$

In the performance index (8.25), the matrices are:

$$Q = \begin{pmatrix} 0 & 0 \\ 0 & 4 \end{pmatrix}, \qquad R = 1$$

The terminal time is $T = \pi/2$.

Set the number of block pulse functions to $m = 4$. Since the matrix F in the canonical equation is:

$$F = \begin{pmatrix} 0 & 0 & 1 & 0 \\ 1 & 0 & 0 & 0 \\ 0 & 0 & 0 & -1 \\ 0 & 4 & 0 & 0 \end{pmatrix}$$

and the matrices Γ and Λ are:

$$\Gamma = \left(I - \frac{h}{2}F \right)^{-1}$$

and

$$\Lambda = \Gamma \left(I + \frac{h}{2}F \right)$$

respectively, the block pulse coefficients of the transition matrix can be computed iteratively from (8.16) and (8.38), i.e. $\Psi_4 = \Gamma$, $\Psi_3 = \Lambda\Psi_4$, $\Psi_2 = \Lambda\Psi_3$ and $\Psi_1 = \Lambda\Psi_2$. Based on the block pulse coefficients of parts of the transition matrix:

$$\begin{aligned}
\Psi_{21}(T,t) &= \begin{pmatrix} -1.8063 & -3.5180 \\ 3.5180 & 4.4508 \end{pmatrix} \phi_1(t) + \begin{pmatrix} -0.7369 & -1.9280 \\ 1.9280 & 3.6471 \end{pmatrix} \phi_2(t) \\
&+ \begin{pmatrix} -0.2093 & -0.7593 \\ 0.7593 & 2.3054 \end{pmatrix} \phi_3(t) + \begin{pmatrix} -0.0301 & -0.1533 \\ 0.1533 & 0.7808 \end{pmatrix} \phi_4(t)
\end{aligned}$$

and

$$\begin{aligned}
\Psi_{22}(T,t) &= \begin{pmatrix} 0.2619 & -1.1127 \\ 1.8063 & 0.2619 \end{pmatrix} \phi_1(t) + \begin{pmatrix} 0.7613 & -0.9118 \\ 0.7369 & 0.7613 \end{pmatrix} \phi_2(t) \\
&+ \begin{pmatrix} 0.9471 & -0.5763 \\ 0.2093 & 0.9471 \end{pmatrix} \phi_3(t) + \begin{pmatrix} 0.9941 & -0.1952 \\ 0.0301 & 0.9941 \end{pmatrix} \phi_4(t)
\end{aligned}$$

the piecewise constant feedback gains can be obtained from (8.39):

$$\begin{aligned}
K(t) &= \begin{pmatrix} 1.6558 \\ 1.9394 \end{pmatrix} \phi_1(t) + \begin{pmatrix} 0.9564 \\ 1.4843 \end{pmatrix} \phi_2(t) \\
&+ \begin{pmatrix} 0.2352 \\ 0.5990 \end{pmatrix} \phi_3(t) + \begin{pmatrix} 0.0001 \\ 0.0003 \end{pmatrix} \phi_4(t)
\end{aligned}$$

Theoretically, the optimal time-varying feedback gains are (Chen and Hsiao, 1975):

$$k_1(t) = \frac{\sinh(\pi - 2t) - \sin(\pi - 2t)}{\cosh^2(\pi/2 - t) + \cos^2(\pi/2 - t)}$$

and

$$k_2(t) = \frac{\cosh(\pi - 2t) - \cos(\pi - 2t)}{\cosh^2(\pi/2 - t) + \cos^2(\pi/2 - t)}$$

This optimal feedback gain and its piecewise constant approximation obtained from the above block pulse function method with $m = 16$ are illustrated in Figure 8.3 for comparison.

The block pulse functions can also be applied to solve the Riccati differential equation for determining the feedback gains of the same optimal control problem. As is well known, the time-varying gain matrix $K(t)$ can be determined from:

$$K(t) = R^{-1}B^T L(t) \tag{8.40}$$

where $L(t)$ is an n-dimensional symmetrical matrix and satisfies the Riccati equation:

$$\dot{L}(t) = V(t) \tag{8.41}$$

with $L(T) = 0$ and

$$V(t) = -L(t)A - A^T L(t) + L(t)BR^{-1}B^T L(t) - Q \tag{8.42}$$

Integrating (8.41) backwards from T to t on both sides, we have:

$$L(t) = \int_T^t V(t)dt \tag{8.43}$$

According to the backward integral operation rule (2.92), the block pulse series of (8.43) is:

$$\sum_{i=1}^{m} L_i\phi_i(t) \doteq \sum_{i=1}^{m} \left(-\frac{h}{2}V_i - h\sum_{j=i+1}^{m} V_j \right) \phi_i(t) \tag{8.44}$$

Equating the coefficients of each block pulse function $\phi_i(t)$ $(i = 1, 2, \ldots, m)$ in the above equation separately, we can obtain m equations of block pulse coefficients. If we denote

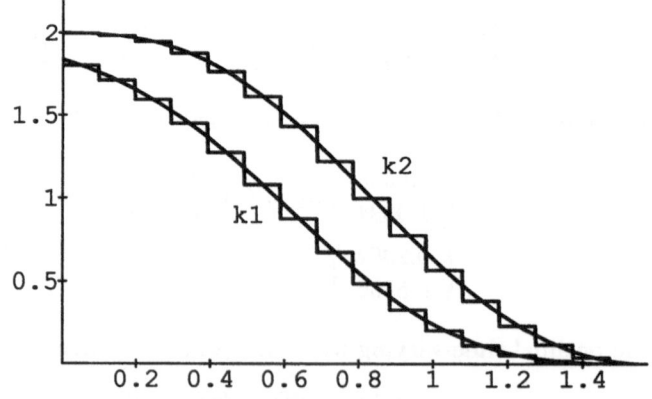

Figure 8.3: Feedback gains in optimal control of a time-invariant linear system.

the ith equation as $E_{(i)}$, and apply the operation $E_{(i+1)} - E_{(i)}$ $(i = 1, 2, \ldots, m - 1)$ on the two neighbouring equations, we can obtain:

$$
L_i = \begin{cases} -\dfrac{h}{2} V_i & \text{for } i = m \\[2mm] L_{i+1} - \dfrac{h}{2}(V_i + V_{i+1}) & \text{for } i = m-1, m-2, \ldots, 1 \end{cases} \tag{8.45}
$$

where

$$
V_i = -L_i A - A^T L_i + L_i B R^{-1} B^T L_i - Q \tag{8.46}
$$

Thus, the solution of the Riccati differential equation (8.41) is reduced to the solution of a set of nonlinear algebraic equations (8.45).

Discussions about block pulse function methods for solving optimal control problems of time-invariant systems can be found in the papers of Rao and Rao (1979), Shih (1981).

8.2 Time-varying linear systems

Consider a time-varying linear system which is described by the state equation:

$$
\dot{X}(t) = A(t)X(t) + B(t)U(t) \tag{8.47}
$$

and the output equation:

$$
Y(t) = C(t)X(t) + D(t)U(t) \tag{8.48}
$$

where $U(t)$, $X(t)$ and $Y(t)$ are r-, n- and p-dimensional vectors of input, state and output, and $A(t)$, $B(t)$, $C(t)$ and $D(t)$ are time-varying matrices of appropriate dimensions, respectively.

8.2.1 System analysis

Similar to the case of time-invariant linear systems, the main analysis problem here is to evaluate the block pulse coefficients of the states $X(t)$ from the system description (8.47) and the initial value of the states $X(0)$ under the input excitations $U(t)$ in a finite interval $t \in [0, T]$. For this purpose, we also integrate the state equation (8.47) from 0 to t on both sides:

$$
X(t) - X(0) = \int_0^t \Big(A(t)X(t) + B(t)U(t) \Big) dt \tag{8.49}
$$

and expand it into block pulse series:

$$
\sum_{k=1}^{m} (X_k - X(0))\, \phi_k(t) \doteq \sum_{k=1}^{m} (A_k X_k + B_k U_k) \left(\frac{h}{2}\phi_k(t) + h \sum_{j=k+1}^{m} \phi_j(t) \right) \tag{8.50}
$$

Equating the coefficients of each block pulse function $\phi_k(t)$ $(k = 1, 2, \ldots, m)$ in the above equation separately, we can obtain m equations, e.g. the kth equation $E_{(k)}$ is :

$$X_k - X(0) = \frac{h}{2}(A_k X_k + B_k U_k) + h \sum_{j=1}^{k-1}(A_j X_j + B_j U_j) \qquad (8.51)$$

Applying the operation $E_{(k+1)} - E_{(k)}$ on the two neighbouring equations, we have:

$$X_{k+1} - X_k = \frac{h}{2}(A_{k+1} X_{k+1} + A_k X_k + B_{k+1} U_{k+1} + B_k U_k) \qquad (8.52)$$

where $k = 1, 2, \ldots, m-1$. Starting from

$$X_1 = X(0) + \frac{h}{2}(A_1 X_1 + B_1 U_1) \qquad (8.53)$$

the block pulse coefficients of the states can be evaluated recursively by (8.52), i.e. we first compute X_1 as the initial value of the recursion from:

$$X_1 = \left(I - \frac{h}{2}A_1\right)^{-1}\left(X(0) + \frac{h}{2}B_1 U_1\right) \qquad (8.54)$$

and then compute X_{k+1} $(k = 1, 2, \ldots, m-1)$ further from:

$$X_{k+1} = \left(I - \frac{h}{2}A_{k+1}\right)^{-1}\left(\left(I + \frac{h}{2}A_k\right)X_k + \frac{h}{2}(B_{k+1} U_{k+1} + B_k U_k)\right) \qquad (8.55)$$

Since (8.55) expresses the relation between the input and state vectors of a time-varying linear system in a piecewise constant approximate way, it can also be regarded as the block pulse difference equation of (8.47).

Example 8.5 Determine the block pulse coefficients of the state variables of a time-varying linear system described by (8.47) in the interval $t \in [0, 1)$ with $m = 10$. Here, the system matrix is:

$$A(t) = \begin{pmatrix} \cos t & \sin t \\ -\sin t & \cos t \end{pmatrix}$$

and no external excitations exist. The initial values of the states are $x_1(0) = 1$, $x_2(0) = 2$.

To obtain the block pulse coefficients of the state vector $X(t)$, the matrix $A(t)$ is first expanded into its block pulse series. After the first block pulse coefficient X_1 is computed by (8.54), the remaining block pulse coefficients X_2, X_3, ... can be computed recursively by (8.55). We can notice that matrix inversion is involved in each step of the recursion. The piecewise constant approximate solutions of the state variables thus obtained are:

$$X(t) \doteq \begin{pmatrix} 1.0581 \\ 2.1023 \end{pmatrix}\phi_1(t) + \begin{pmatrix} 1.1923 \\ 2.3098 \end{pmatrix}\phi_2(t)$$

$$+ \begin{pmatrix} 1.3658 \\ 2.5202 \end{pmatrix}\phi_3(t) + \begin{pmatrix} 1.5842 \\ 2.7262 \end{pmatrix}\phi_4(t) + \cdots$$

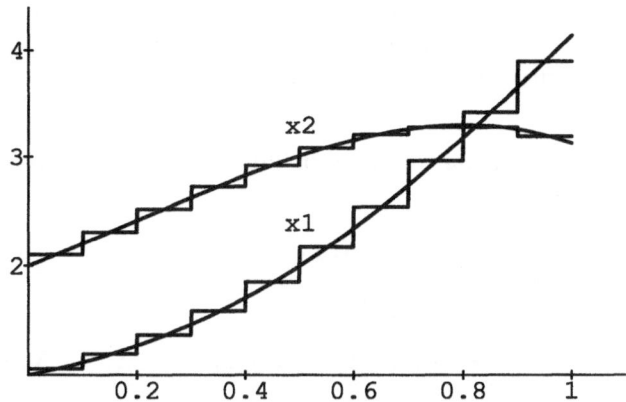

Figure 8.4: State variables of a time-varying linear system.

For comparison, they are illustrated in Figure 8.4 together with their analytical solutions (Wu and Sherif, 1976):

$$x_1(t) = \Big(\cos(1 - \cos t) + 2\sin(1 - \cos t) \Big) e^{\sin t}$$

and

$$x_2(t) = \Big(-\sin(1 - \cos t) + 2\cos(1 - \cos t) \Big) e^{\sin t}$$

Like the discussion of time-invariant linear systems in Section 8.1, the transition matrix of a time-varying linear system can also be expressed by its block pulse series approximation:

$$\Psi(t, 0) \doteq \sum_{k=1}^{m} R_k \phi(t) \tag{8.56}$$

Set $B(t) = 0$, we can obtain from (8.54) and (8.55):

$$R_1 = \left(I - \frac{h}{2} A_1 \right)^{-1} \tag{8.57}$$

and

$$R_{k+1} = \left(I - \frac{h}{2} A_{k+1} \right)^{-1} \left(I + \frac{h}{2} A_k \right) R_k \tag{8.58}$$

where $k = 1, 2, \ldots, m - 1$. Using these relations, we can compute the block pulse coefficients of a transition matrix of a time-varying linear system iteratively. But unlike the time-invariant case, matrix inversion must be done in every step of the computation.

8.2.2 System identification

For continuous time-varying linear systems, the identification problem in the block pulse domain is to estimate the block pulse coefficients of the parameter matrices $A(t)$

and $B(t)$ in (8.47). Since each entry in these matrices contains m unknown block pulse coefficients which should be identified, at least $(n + r)$ independent input signals should be used to ensure the uniqueness of the block pulse coefficients of the estimated parameters.

Suppose that q input signals are used in this identification problem. We first define the $q \times n$ matrix F_1 as:

$$F_1^T = \left(\begin{array}{cccc} X_{1,1} - X_1(0) & X_{2,1} - X_2(0) & \cdots & X_{q,1} - X_q(0) \end{array} \right) \tag{8.59}$$

the $q \times n$ matrices F_k $(k = 2, 3, \ldots, m)$:

$$F_k^T = \left(\begin{array}{cccc} X_{1,k} - X_{1,k-1} & X_{2,k} - X_{2,k-1} & \cdots & X_{q,k} - X_{q,k-1} \end{array} \right) \tag{8.60}$$

the $q \times (n + r)$ matrices G_k $(k = 1, 2, \ldots, m)$:

$$G_k^T = \frac{h}{2} \left(\begin{array}{cccc} X_{1,k} & X_{2,k} & \cdots & X_{q,k} \\ U_{1,k} & U_{2,k} & \cdots & U_{q,k} \end{array} \right) \tag{8.61}$$

and the $(n + r) \times n$ matrices Θ_k $(k = 1, 2, \ldots, m)$:

$$\Theta_k^T = \left(\begin{array}{cc} A_k & B_k \end{array} \right) \tag{8.62}$$

Then we can rearrange (8.53) and (8.52) in the form:

$$F_k = \begin{cases} G_1 \Theta_1 & \text{for } k = 1 \\ G_k \Theta_k + G_{k-1} \Theta_{k-1} & \text{for } k = 2, 3, \ldots, m \end{cases} \tag{8.63}$$

When $q \geq n + r$, the unknown value of Θ_k can be estimated successively, e.g. the ordinary least-squares algorithm gives:

$$\hat{\Theta}_k = \begin{cases} \left(G_1^T G_1 \right)^{-1} G_1^T F_1 & \text{for } k = 1 \\ \left(G_k^T G_k \right)^{-1} G_k^T \left(F_k - G_{k-1} \hat{\Theta}_{k-1} \right) & \text{for } k = 2, 3, \ldots, m \end{cases} \tag{8.64}$$

where $\hat{\Theta}_{k-1}$ and $\hat{\Theta}_k$ are the estimated values of the matrices Θ_{k-1} and Θ_k, respectively.

Example 8.6 Consider a time-varying linear system with $n = 1$ described by (8.47). The system parameters are $a(t) = -e^{-t}$ and $b(t) = e^{-2t}$. Four input signals are used in this identification problem. They are $u_1(t) = e^{-2t}$, $u_2(t) = 2t$, $u_3(t) = te^{-t}$ and $u_4(t) = 1/(t + 0.5)$. Our problem is to estimate the unknown parameters from the simulated data of input and states in the interval $t \in [0, 1)$. Here the sampling period is $h = 0.05$.

Based on the block pulse coefficients of the four input and output signals, the matrices F_k and G_k $(k = 1, 2, \ldots, m)$ can first be constructed, and estimations can then be done by using (8.64). The estimated results are:

$$\hat{a}(t) = \left(\begin{array}{ccccc} -0.9752 & -0.9276 & -0.8824 & -0.8394 & \cdots \end{array} \right) \Phi(t)$$

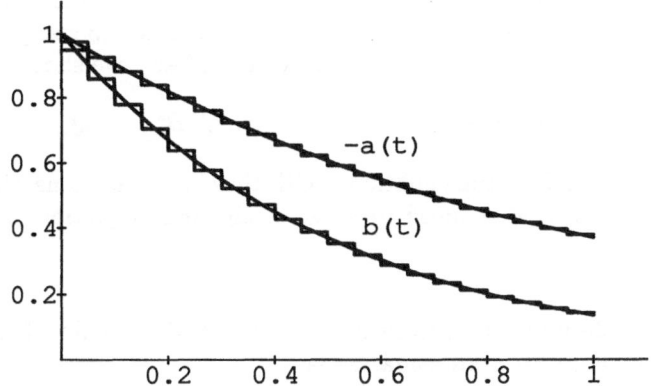

Figure 8.5: Parameter estimation of a time-varying linear system.

and

$$\hat{b}(t) = \left(\begin{array}{ccccc} 0.9503 & 0.8600 & 0.7783 & 0.7043 & \cdots \end{array} \right) \Phi(t)$$

which are illustrated in Figure 8.5 together with the true time-varying parameters. Discussions about block pulse function methods for solving analysis and identification problems of time-varying linear systems can be found in the papers of Shih (1978), Rao and Ganapathy (1979), Chen and Meng (1982), Jaw and Kung (1984).

8.2.3 Optimal control

Consider an optimal control problem of the time-varying linear system (8.47) with respect to a quadratic performance index over a finite interval of time:

$$J = \frac{1}{2} \int_0^T \left(X^T(t)Q(t)X(t) + U^T(t)R(t)U(t) \right) dt \tag{8.65}$$

where $Q(t)$ is a positive semi-definite n-dimensional matrix and $R(t)$ is a positive definite r-dimensional matrix. The solution of this optimal problem is similar to that of time-invariant linear systems discussed in Section 8.1, only the matrices of constants there should be modified to the matrices of time functions to suit the time-varying case. For example, the equations (8.28), (8.33) and (8.42) are now modified respectively to:

$$F = \left(\begin{array}{cc} A(t) & B(t)R^{-1}(t)B^T(t) \\ Q(t) & -A^T(t) \end{array} \right) \tag{8.66}$$

$$K(t) = R^{-1}(t)B^T(t)\Psi_{22}^{-1}(T,t)\Psi_{21}(T,t) \tag{8.67}$$

and

$$V(t) = -L(t)A(t) - A^T(t)L(t) + L(t)B(t)R^{-1}(t)B^T(t)L(t) - Q(t) \tag{8.68}$$

Correspondingly, the block pulse function method in Section 8.1 should also be modified. Instead of (8.16) and (8.9), we should use (8.57) and (8.58) to compute the block pulse coefficients of the transition matrix, and instead of (8.46), we should use the relation:

$$V_i = -L_i A_i - A_i^T L_i + L_i B_i R_i^{-1} B_i^T L_i - Q_i \qquad (8.69)$$

to solve the Riccati differential equation. With these modifications, the optimal control problem here can also be reduced to a set of algebraic equations in the block pulse domain.

Example 8.7 Determine the piecewise constant feedback gain of the above optimal control problem. Here the state equation is:

$$\dot{X}(t) = t X(t) + U(t)$$

and the performance index is:

$$J = \frac{1}{2} \int_0^T \left(X^2(t) + U^2(t) \right) dt$$

The terminal time is $T = 1$.

Set the number of block pulse functions to $m = 4$. Since the matrix F in the canonical equation is:

$$F(t) = \begin{pmatrix} t & 1 \\ 1 & -t \end{pmatrix}$$

the block pulse coefficients of the transition matrix can be computed successively from (8.57) and (8.58), i.e.

$$\begin{aligned}
\Psi_4 &= \left(I - \frac{1}{8} F_4 \right)^{-1} \\
\Psi_3 &= \left(I - \frac{1}{8} F_3 \right)^{-1} \left(I + \frac{1}{8} F_4 \right) \Psi_4 \\
\Psi_2 &= \left(I - \frac{1}{8} F_2 \right)^{-1} \left(I + \frac{1}{8} F_3 \right) \Psi_3 \\
\Psi_1 &= \left(I - \frac{1}{8} F_1 \right)^{-1} \left(I + \frac{1}{8} F_2 \right) \Psi_2
\end{aligned}$$

From the parts of the transition matrix:

$$\Psi_{21}(T,t) \doteq \left(\begin{array}{cccc} 0.9331 & 0.6586 & 0.3896 & 0.1286 \end{array} \right) \Phi(t)$$

and

$$\Psi_{22}(T,t) \doteq \left(\begin{array}{cccc} 0.9883 & 0.8443 & 0.8167 & 0.9159 \end{array} \right) \Phi(t)$$

the piecewise constant feedback gain can be obtained from (8.67):

$$K(t) \doteq \left(\begin{array}{cccc} 0.9442 & 0.7800 & 0.4771 & 0.1404 \end{array} \right) \Phi(t) \qquad (8.70)$$

In the above solution, we can notice that matrix inversions are involved in every step. Therefore computations will increase in solving problems of time-varying systems, especially when the system order n is large. The piecewise constant solution with $m = 20$ obtained from the same procedure is illustrated in Figure 8.6 together with the exact optimal feedback gain.

If we solve this optimal problem from the Riccati differential equation, we should solve the nonlinear differential equation:

$$\dot{L} = -2tL + L^2 - 1$$

But in the block pulse domain, the nonlinear differential equation is reduced to some nonlinear algebraic equations. According to (8.45) and (8.69), we can first obtain from the equation:

$$L_4 = -\frac{1}{8}\left(-\frac{1}{4}L_4 + L_4^2 - 1\right)$$

the value $L_4 = 0.1561$, and then from the equation:

$$L_3 = -\frac{1}{8}\left(-\frac{3}{4}L_3 + L_3^2 - 1 + V_4\right)$$

the value $L_3 = 0.4835$, and further $L_2 = 0.7772$ and $L_1 = 0.9429$. These values are just the block pulse coefficients of $K(t)$. In fact, the piecewise constant solutions of this optimal control problem from the above two different block pulse function methods are very close. For example, with the number of block pulse functions $m = 20$, the one obtained from solving the transition matrix is:

$$K(t) \doteq \Big(\begin{array}{ccccc} 0.9663 & 0.9579 & 0.9436 & 0.9232 & 0.8965 \\ 0.8633 & 0.8236 & 0.7777 & 0.7259 & 0.6689 \\ 0.6074 & 0.5424 & 0.4748 & 0.4059 & 0.3369 \\ 0.2688 & 0.2027 & 0.1397 & 0.0805 & 0.0256 \end{array} \Big) \Phi(t)$$

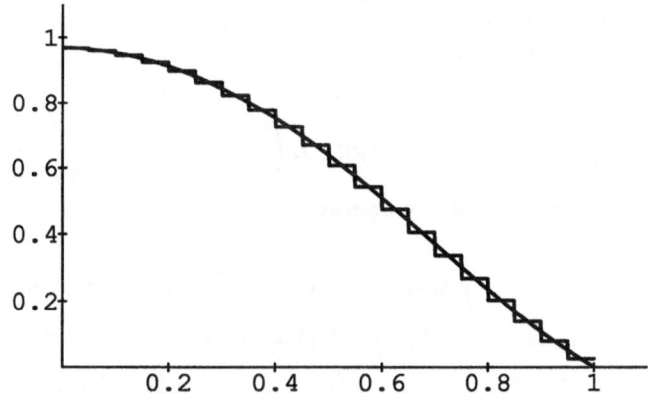

Figure 8.6: Feedback gain in optimal control of a time-varying linear system.

and the one obtained from solving the Riccati differential equation is:

$$K(t) \doteq \big(\ \ 0.9664 \quad 0.9579 \quad 0.9436 \quad 0.9232 \quad 0.8965$$

$$0.8632 \quad 0.8235 \quad 0.7776 \quad 0.7259 \quad 0.6689$$

$$0.6075 \quad 0.5425 \quad 0.4751 \quad 0.4063 \quad 0.3373$$

$$0.2693 \quad 0.2034 \quad 0.1403 \quad 0.0811 \quad 0.0263 \ \big) \ \Phi(t)$$

Discussions about block pulse function methods for solving optimal control problems of time-varying systems can be found in the papers of Shih (1981), Hsu and Cheng (1981), Xu and Zheng (1982), Hwang and Shih (1985), Xing and Wang (1985), Hwang, Shih and Kung (1986).

8.3 Linear systems containing time delays

Containing time delay τ in the input and states, a time-invariant linear system can be described by:

$$\dot{X}(t) = AX(t) + BU(t) + LX(t - \tau) + MU(t - \tau) \tag{8.71}$$

and a time-varying linear system can be described by:

$$\dot{X}(t) = A(t)X(t) + B(t)U(t) + L(t)X(t - \tau) + M(t)U(t - \tau) \tag{8.72}$$

In the above descriptions, $U(t)$ and $X(t)$ are r- and n-dimensional vectors of input and state, respectively. The delay time is $\tau = (q + \lambda)h$ with an integer $q \geq 0$ and a fraction $0 \leq \lambda < 1$.

8.3.1 System analysis

The analysis problem using the block pulse functions is to evaluate the block pulse coefficients of the state vector in a finite interval $t \in [0, T)$ from the state equation and the known value of the states for $t \leq 0$ under the input excitations. Here we consider the time-invariant case (8.71). If $X(t) = 0$ and $U(t) = 0$ for $t < 0$, we can first integrate the state equation from 0 to t on both sides:

$$X(t) - X(0) = A\int_0^t X(t)dt + B\int_0^t U(t)dt + L\int_0^t X(t - \tau)dt + M\int_0^t U(t - \tau)dt \tag{8.73}$$

and then use the integral and delay operation rules of block pulse series to obtain the equation:

$$\big(\ X_1 \quad X_2 \quad \cdots \quad X_m \ \big) \Phi(t) - \big(\ X(0) \quad X(0) \quad \cdots \quad X(0) \ \big) \Phi(t)$$

$$\doteq A\big(\ X_1 \quad X_2 \quad \cdots \quad X_m \ \big) P\Phi(t) + B\big(\ U_1 \quad U_2 \quad \cdots \quad U_m \ \big) P\Phi(t)$$

$$+ L\big(\ X_1 \quad X_2 \quad \cdots \quad X_m \ \big) \big((1 - \lambda)H^q + \lambda H^{q+1}\big) P\Phi(t)$$

$$+ M\big(\ U_1 \quad U_2 \quad \cdots \quad U_m \ \big) \big((1 - \lambda)H^q + \lambda H^{q+1}\big) P\Phi(t) \tag{8.74}$$

Since both the conventional integration operational matrix P and the delay matrix $(1-\lambda)H^q + \lambda H^{q+1}$ have special forms, we can obtain m equations of block pulse coefficients $E_{(k)}$ by equating the block pulse coefficients of $\phi_k(t)$ ($k = 1, 2, \ldots, m$) on both sides of the above equation separately. These equations can be divided into three cases. For $k = 1, 2, \ldots, q$, the equations $E_{(k)}$ are:

$$X_k - X(0) = \frac{h}{2}(AX_k + BU_k) + h\sum_{j=1}^{k-1}(AX_j + BU_j) \tag{8.75}$$

for $k = q + 1$, the equation $E_{(k)}$ is:

$$X_k - X(0) = \frac{h}{2}(AX_k + BU_k) + h\sum_{j=1}^{k-1}(AX_j + BU_j) + \frac{h}{2}(1-\lambda)(LX_{k-q} + MU_{k-q}) \tag{8.76}$$

and for $k = q + 2, q + 3, \ldots, m$, the equations $E_{(k)}$ are:

$$X_k - X(0) = \frac{h}{2}(AX_k + BU_k) + h\sum_{j=1}^{k-1}(AX_j + BU_j)$$
$$+ \frac{h}{2}(1-\lambda)(LX_{k-q} + MU_{k-q}) + \frac{h}{2}\lambda(LX_{k-q-1} + MU_{k-q-1})$$
$$+ h(1-\lambda)\sum_{j=1}^{k-q-1}(LX_j + MU_j) + h\lambda\sum_{j=1}^{k-q-2}(LX_j + MU_j) \tag{8.77}$$

From these equations, the block pulse coefficients of the state vector can be evaluated successively.

We notice that the block pulse coefficients of $X(t)$ and $U(t)$ in some previous subintervals are involved in (8.76) and (8.77) because time delay exists in certain terms of states and inputs, and we also notice that some terms are omitted in (8.75) and (8.76) because the special conditions are $X(t) = 0$ and $U(t) = 0$ for $t < 0$. In fact, (8.77) can be used to express all the m equations of block pulse coefficients for all the three cases above where $X(t)$ and $U(t)$ are known for $t < 0$, only nonpositive integers should be used as subscripts to denote the block pulse coefficients in the interval $t \in [-(q+1)h, 0)$.

Similar to the discussion in Section 7.1, the algorithm (8.77) for solving the system analysis problem has a disadvantage because the size of computation will increase when the number of block pulse functions m is large. In order to avoid this disadvantage, we also apply the operation $E_{(k+1)} - E_{(k)}$ on the two neighbouring equations to obtain:

$$X_{k+1} - X_k = \frac{h}{2}A(X_{k+1} + X_k) + \frac{h}{2}B(U_{k+1} + U_k)$$
$$+ \frac{h}{2}(1-\lambda)L(X_{k-q+1} + X_{k-q}) + \frac{h}{2}\lambda L(X_{k-q} + X_{k-q-1})$$
$$+ \frac{h}{2}(1-\lambda)M(U_{k-q+1} + U_{k-q}) + \frac{h}{2}\lambda M(U_{k-q} + U_{k-q-1}) \tag{8.78}$$

With X_1 as the starting value of the recursion:

$$X_1 = \Gamma\left(X(0) + \frac{h}{2}(BU_1 + LX_{1-q} + MU_{1-q})\right) \tag{8.79}$$

the block pulse coefficients of the state vector $X(t)$ in (8.71) can be computed recursively from the block pulse difference equation:

$$X_{k+1} = \Gamma\left(\left(I + \frac{h}{2}A\right)X_k + \frac{h}{2}B\left(U_{k+1} + U_k\right)\right.$$

$$+\frac{h}{2}(1-\lambda)L\left(X_{k-q+1} + X_{k-q}\right) + \frac{h}{2}\lambda L\left(X_{k-q} + X_{k-q-1}\right)$$

$$\left.+\frac{h}{2}(1-\lambda)M\left(U_{k-q+1} + U_{k-q}\right) + \frac{h}{2}\lambda M\left(U_{k-q} + U_{k-q-1}\right)\right) \qquad (8.80)$$

where $k = 1, 2, \ldots, m - 1$ and $q > 0$.

In order to get satisfactory approximation of the results, we usually set the width of the block pulses h small enough. In such cases, it is also reasonable to assume that the delay time is qh, i.e. the fraction part $\lambda h = 0$. Under this assumption, the expressions in the above discussion can be simplified greatly. For example, (8.78) becomes:

$$X_{k+1} - X_k = \frac{h}{2}A\left(X_{k+1} + X_k\right) + \frac{h}{2}B\left(U_{k+1} + U_k\right)$$

$$+\frac{h}{2}L\left(X_{k-q+1} + X_{k-q}\right) + \frac{h}{2}M\left(U_{k-q+1} + U_{k-q}\right) \qquad (8.81)$$

and (8.80) becomes:

$$X_{k+1} = \Gamma\left(\left(I + \frac{h}{2}A\right)X_k + \frac{h}{2}B\left(U_{k+1} + U_k\right)\right.$$

$$\left.+\frac{h}{2}L\left(X_{k-q+1} + X_{k-q}\right) + \frac{h}{2}M\left(U_{k-q+1} + U_{k-q}\right)\right) \qquad (8.82)$$

Example 8.8 Determine the block pulse coefficients of the state variables of a linear system described by (8.71) in the interval $t \in [0, 6)$ with $m = 18$. Here, the parameter matrices are

$$A = \begin{pmatrix} 0 & 1 \\ 0 & 0 \end{pmatrix}, \quad B = \begin{pmatrix} 0 \\ 0 \end{pmatrix}, \quad L = \begin{pmatrix} 0 & 0 \\ -0.3 & -1 \end{pmatrix}, \quad M = \begin{pmatrix} 0 \\ 1 \end{pmatrix}$$

with the delay time $\tau = 1.0$. The value of the states are zeros for $t \le 0$, and the input excitation is the composition of a step function and a ramp function $u(t) = (1+0.3t)\mu(t)$.

The input signal is first expanded into its block pulse series, and then the block pulse series approximation of $X(t)$ can be computed recursively from (8.79) and (8.80):

$$X(t) \doteq \begin{pmatrix} 0.0292 \\ 0.1750 \end{pmatrix}\phi_4(t) + \begin{pmatrix} 0.1486 \\ 0.5417 \end{pmatrix}\phi_5(t)$$

$$+ \begin{pmatrix} 0.3958 \\ 0.9417 \end{pmatrix}\phi_6(t) + \begin{pmatrix} 0.7768 \\ 1.3444 \end{pmatrix}\phi_7(t) + \cdots$$

The analytical solutions of the state variables and their piecewise constant approximations are illustrated in Figure 8.7 for comparison.

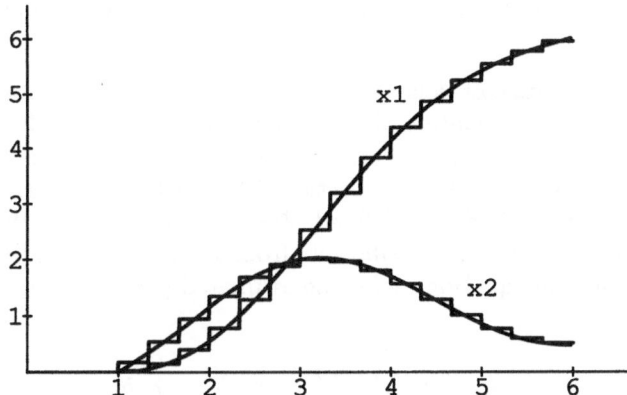

Figure 8.7: State variables of a linear system containing time delay.

Combining the discussions of time-invariant delay systems above and time-varying linear systems in Section 8.2, the formulas of the block pulse function method for time-varying delay systems can also be obtained. For example, the equations similar to (8.79), (8.82) are:

$$X_1 = \left(I - \frac{h}{2}A_1\right)^{-1}\left(X(0) + \frac{h}{2}\left(B_1U_1 + L_1X_{1-q} + M_1U_{1-q}\right)\right) \qquad (8.83)$$

$$X_{k+1} = \left(I - \frac{h}{2}A_{k+1}\right)^{-1}\left(\left(I + \frac{h}{2}A_k\right)X_k + \frac{h}{2}\left(B_{k+1}U_{k+1} + B_kU_k\right)\right.$$
$$\left. + \frac{h}{2}\left(L_{k+1}X_{k-q+1} + L_kX_{k-q}\right) + \frac{h}{2}\left(M_{k+1}U_{k-q+1} + M_kU_{k-q}\right)\right) \qquad (8.84)$$

and the relation similar to (8.81) is:

$$X_{k+1} - X_k = \frac{h}{2}\left(A_{k+1}X_{k+1} + A_kX_k\right) + \frac{h}{2}\left(B_{k+1}U_{k+1} + B_kU_k\right)$$
$$+ \frac{h}{2}\left(L_{k+1}X_{k-q+1} + L_kX_{k-q}\right) + \frac{h}{2}\left(M_{k+1}U_{k-q+1} + M_kU_{k-q}\right) \qquad (8.85)$$

The procedures for analysis problems of time-varying delay systems are almost the same as before, only matrix inversion should be done in every step of the computation.

8.3.2 System identification

For both time-invariant and time-varying delay systems, our problem is to estimate the unknown parameter matrices from the block pulse coefficients of the inputs $U(t)$ and the states $X(t)$. If the delay time τ is unknown, it should also be estimated together with the parameter matrices. Since we have already obtained the relations between inputs and states in the block pulse domain, this identification problem can be solved easily.

Consider a time-invariant delay system (8.71). If the delay time $\tau = qh$ is known, we can directly use the estimation algorithms (8.21) or (8.24), only the matrices F, G, Θ or the vectors f_k, g_k should be reconstructed. If the delay time is unknown, we can also use the procedures discussed in Section 7.4 to detect the correct delay time, but now the loss functions should be redefined. For example, if we use the equation (8.81) p times with different block pulse coefficients from the measured data, and if we express the error vector of each equation under the estimated parameter matrices as $e_j(\tau)$ $(j = j_1, j_2, \ldots, j_p)$:

$$
\begin{aligned}
e_j(\tau) = {} & X_{k+1} - \left(I - \frac{h}{2}\hat{A}\right)^{-1} \left(\left(I + \frac{h}{2}\hat{A}\right) X_k + \frac{h}{2}\hat{B}\left(U_{k+1} + U_k\right)\right. \\
& \left. + \frac{h}{2}\hat{L}\left(X_{k-q+1} + X_{k-q}\right) + \frac{h}{2}\hat{M}\left(U_{k-q+1} + U_{k-q}\right)\right)
\end{aligned}
\tag{8.86}
$$

we can define a loss function:

$$
J(\tau) = \frac{1}{p}\sum_{j=1}^{p} e_j^T(\tau)e_j(\tau)
\tag{8.87}
$$

to indicate the correct delay time, i.e. repeating the estimation procedure as τ increases each time by h until the loss function $J(\tau)$ reaches its minimum. Thus, the correct estimations of the parameter matrices and the delay time can be obtained.

Example 8.9 Consider a time-invariant delay system which is the same as in Example 8.8. Here the input is also the composition of a step function and a ramp function $u(t) = (1 + 0.3t)\mu(t)$. Our problem is to estimate the unknown delay time and parameter matrices from the simulated data of input and states in the interval $t \in [0, 8)$. The sampling period is $h = 0.05$.

For this identification problem, we use the relation (8.81) and the loss function (8.87) with $p = 130$. Under different delay times, the parameter matrices are estimated by the ordinary least-squares algorithm (8.21). The estimation results and the corresponding values of the loss function are listed in Table 8.2. It is obvious that under the delay time $\tau = 1.0$, the parameter matrices have their correct values.

The identification problem of time-varying delay systems in the block pulse domain can be solved based on the relation (8.85). Similar to the identification of time-varying linear systems discussed in Section 8.2, enough independent input signals should be used to ensure the uniqueness of the estimated block pulse coefficients of the parameter matrices. Discussions about block pulse function methods for solving analysis and identification problems of linear systems containing time delays can be found in the papers of Chen (1983), Rao and Srinivasan (1978a), Chen and Jeng (1981), Shih, Hwang and Chia (1980).

$\hat{\tau}$	0.90	0.95	1.00	1.05	1.10	True value
$\hat{a}_{11}$	0.0004	0.0003	0.0000	-0.0001	0.0001	0.0
$\hat{a}_{12}$	0.9997	0.9998	1.0000	1.0000	0.9999	1.0
$\hat{a}_{21}$	-0.6180	-0.1798	0.0003	-0.2868	-0.9631	0.0
$\hat{a}_{22}$	0.6192	0.2240	-0.0002	0.1549	0.6751	0.0
$\hat{b}_1$	-0.0005	-0.0003	0.0000	0.0000	-0.0001	0.0
$\hat{b}_2$	0.6637	0.1244	-0.0002	0.4746	1.2052	0.0
$\hat{l}_{11}$	-0.0005	-0.0003	0.0000	0.0001	-0.0001	0.0
$\hat{l}_{12}$	-0.0002	-0.0001	0.0000	0.0000	0.0000	0.0
$\hat{l}_{21}$	0.4367	-0.0794	-0.3003	0.0274	0.8121	-0.3
$\hat{l}_{22}$	-0.8020	-0.9432	-1.0001	-0.8322	-0.4211	-1.0
$\hat{m}_1$	0.0007	0.0005	0.0000	0.0000	0.0002	0.0
$\hat{m}_2$	-0.0737	0.7367	1.0003	0.3779	-0.7339	1.0
$J(\hat{\tau}) \times 10^{-9}$	20646	6183.7	4.1132	6294.9	18340	

Table 8.2: Estimated parameters under different delay times.

8.4   Bilinear systems

A time-invariant bilinear system can be described by:

$$\dot{X}(t) = AX(t) + BU(t) + \sum_{i=1}^{r} N_i X(t) u_i(t) \tag{8.88}$$

and a time-varying bilinear system can be described by:

$$\dot{X}(t) = A(t)X(t) + B(t)U(t) + \sum_{i=1}^{r} N_i(t) X(t) u_i(t) \tag{8.89}$$

In the above descriptions, $X(t)$ is n-dimensional state vector, $U(t)$ is r-dimensional input vector, and the scalar $u_i(t)$ is the ith entry of $U(t)$. Since bilinear systems are linear both in inputs and in states, they can be manipulated more easily than general nonlinear systems. Due to the simple multiplication rule of block pulse series, if block pulse functions are applied to these systems, their manipulations are almost the same as the ones of linear systems.

8.4.1   System analysis

Our problem here is to evaluate the block pulse coefficients of the states $X(t)$ in a finite interval $t \in [0, T)$ from the state equation and the initial value of the states $X(0)$

under the input excitations $U(t)$. Here we consider the time-invariant case (8.88). After integrating the state equation from 0 to t on both sides:

$$X(t) - X(0) = A \int_0^t X(t)dt + B \int_0^t U(t)dt + \sum_{i=1}^r \left(N_i \int_0^t X(t)u_i(t)dt \right) \qquad (8.90)$$

we obtain the equation based on the operation rules of block pulse series:

$$\begin{aligned}
\left(\begin{array}{cccc} X_1 & X_2 & \cdots & X_m \end{array} \right) &\Phi(t) - \left(\begin{array}{cccc} X(0) & X(0) & \cdots & X(0) \end{array} \right) \Phi(t) \\
\doteq\ & A \left(\begin{array}{cccc} X_1 & X_2 & \cdots & X_m \end{array} \right) P\Phi(t) + B \left(\begin{array}{cccc} U_1 & U_2 & \cdots & U_m \end{array} \right) P\Phi(t) \\
&+ \sum_{i=1}^r N_i \left(\begin{array}{cccc} X_1 u_{i,1} & X_2 u_{i,2} & \cdots & X_m u_{i,m} \end{array} \right) P\Phi(t)
\end{aligned} \qquad (8.91)$$

Due to the special form of the conventional integration operational matrix P, we can obtain m equations of block pulse coefficients $E_{(k)}$ by equating the block pulse coefficients of $\phi_k(t)$ ($k = 1, 2, \ldots, m$) on both sides of the above equation separately:

$$\begin{aligned}
X_k - X(0) &= \frac{h}{2} \left(AX_k + BU_k + \sum_{i=1}^r N_i X_k u_{i,k} \right) \\
&+ h \sum_{j=1}^{k-1} \left(AX_j + BU_j + \sum_{i=1}^r N_i X_j u_{i,j} \right)
\end{aligned} \qquad (8.92)$$

From these equations, the block pulse coefficients of the state vector can be evaluated successively.

As in the discussion in the previous sections, in order to reduce the size of computations, we can apply the operation $E_{(k+1)} - E_{(k)}$ on the two neighbouring equations to obtain:

$$\begin{aligned}
X_{k+1} - X_k &= \frac{h}{2} A \left(X_{k+1} + X_k \right) + \frac{h}{2} B \left(U_{k+1} + U_k \right) \\
&+ \frac{h}{2} \sum_{i=1}^r N_i \left(X_{k+1} u_{i,k+1} + X_k u_{i,k} \right)
\end{aligned} \qquad (8.93)$$

With X_1 as the starting value of the recursion:

$$X_1 = \left(I - \frac{h}{2} A - \frac{h}{2} \sum_{i=1}^r N_i u_{i,1} \right)^{-1} \left(X(0) + \frac{h}{2} BU_1 \right) \qquad (8.94)$$

the block pulse coefficients of the state vector $X(t)$ in (8.88) can be computed recursively from the block pulse difference equation:

$$\begin{aligned}
X_{k+1} = \left(I - \frac{h}{2} A - \frac{h}{2} \sum_{i=1}^r N_i u_{i,k+1} \right)^{-1} &\times \\
\left(\left(I + \frac{h}{2} A + \frac{h}{2} \sum_{i=1}^r N_i u_{i,k} \right) X_k + \frac{h}{2} BU_{k+1} + \frac{h}{2} BU_k \right)
\end{aligned} \qquad (8.95)$$

where $k = 1, 2, \ldots, m - 1$.

Example 8.10 Determine the block pulse coefficients of the state variables of a bilinear system described by (8.88) in the interval $t \in [0, 1)$ with $m = 10$. Here, the parameter matrices are:

$$A = \begin{pmatrix} 0 & 1 \\ -0.5 & -1 \end{pmatrix}, \quad B = \begin{pmatrix} 2 & 1 \\ 1 & 1 \end{pmatrix}, \quad N_1 = \begin{pmatrix} 0 & 1 \\ 0.5 & 0.2 \end{pmatrix}, \quad N_2 = \begin{pmatrix} 1 & 0 \\ 0.2 & 0.3 \end{pmatrix}$$

The initial values of the states are zeros, and the input excitations are $u_1(t) = e^{-0.5t}$ and $u_2(t) = e^{-t}$.

After the input signals are expanded into their block pulse series, the block pulse series approximation of $X(t)$ can be recursively computed from (8.94) and (8.95):

$$X(t) \doteq \left(\begin{matrix} 0.1623 & 0.5089 & 0.8963 & 1.3179 & 1.7664 \\ 0.0953 & 0.2759 & 0.4357 & 0.5737 & 0.6890 \end{matrix} \right.$$

$$\left. \begin{matrix} 2.2345 & 2.7148 & 3.2005 & 3.6848 & 4.1616 \\ 0.7808 & 0.8487 & 0.8926 & 0.9127 & 0.9094 \end{matrix} \right) \Phi(t) \qquad (8.96)$$

The exact solutions of the state variables and their piecewise constant approximations are illustrated in Figure 8.8 for comparison.

The above block pulse function method for the analysis problem of bilinear systems is rather straightforward and is almost the same as the method for linear systems discussed before. But as a disadvantage, matrices which have the form:

$$I - \frac{h}{2}A - \frac{h}{2}\sum_{i=1}^{r} N_i u_{i,k} \qquad (8.97)$$

should be inverted in each step of (8.94) and (8.95). These matrix inversions may cause difficulties when the choice of width of block pulses happens to let the matrices (8.97) be singular. In order to avoid this disadvantage, the Picard's iterative algorithm (Sagaspe, 1976) can be used here to obtain:

$$\lim_{k \to \infty} X^{[k]}(t) = X(t) \qquad (8.98)$$

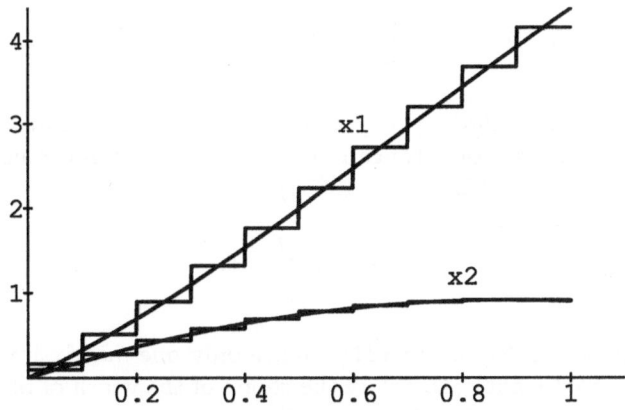

Figure 8.8: State variables of a bilinear system.

where $X^{[k]}(t)$ is an auxiliary vector for iteration:

$$
\dot{X}^{[k]}(t) = \begin{cases} AX^{[1]}(t) + BU(t) & \text{for } k = 1 \\ AX^{[k]}(t) + \sum_{i=1}^{r} N_i X^{[k-1]}(t) u_i(t) & \text{for } k = 2, 3, \ldots \end{cases} \tag{8.99}
$$

with $X^{[k]}(0) = X(0)$. Setting

$$
F^{[k]}(t) = \begin{cases} X^{[1]}(t) & \text{for } k = 1 \\ X^{[k]}(t) - X^{[k-1]}(t) & \text{for } k = 2, 3, \ldots \end{cases} \tag{8.100}
$$

and noticing that the solution of (8.99) can be expressed by the convolution:

$$
X^{[k]}(t) = \begin{cases} e^{At} X(0) + e^{At} * BU(t) & \text{for } k = 1 \\ e^{At} X(0) + e^{At} * \sum_{i=1}^{r} N_i X^{[k-1]}(t) u_i(t) & \text{for } k = 2, 3, \ldots \end{cases} \tag{8.101}
$$

we have:

$$
F^{[k]}(t) = \begin{cases} e^{At} X(0) + e^{At} * BU(t) & \text{for } k = 1 \\ e^{At} * \sum_{i=1}^{r} N_i F^{[k-1]}(t) u_i(t) & \text{for } k = 2, 3, \ldots \end{cases} \tag{8.102}
$$

Using (8.102), (2.59) and (8.16), we can obtain the block pulse coefficients of $F^{[1]}(t)$ as:

$$
F_j^{[1]} = \Lambda^{j-1} \Gamma X(0) + \frac{h}{2} \Lambda^{j-1} \Gamma B U_1 + \frac{h}{2} \sum_{i=1}^{j-1} \Lambda^{i-1} \Gamma B \left(U_{j-i+1} + U_{j-i} \right) \tag{8.103}
$$

and the block pulse coefficients of $F^{[k]}(t)$ as:

$$
F_j^{[k]} = \frac{h}{2} \Lambda^{j-1} \Gamma \sum_{i=1}^{r} N_i F_1^{[k-1]} u_{i,1} + \frac{h}{2} \sum_{l=1}^{j-1} \left(\Lambda^{l-1} \Gamma \sum_{i=1}^{r} N_i \left(F_{j-l+1}^{[k-1]} u_{i,j-l+1} + F_{j-l}^{[k-1]} u_{i,j-l} \right) \right) \tag{8.104}
$$

where $j = 1, 2, \ldots, m$. According to (8.98), satisfactory approximations of block pulse coefficients of the state vector $X(t)$ can be obtained after a finite number of terms are evaluated in the iteration (Marszalek, 1985a):

$$
X_j^{[k]} = \sum_{i=1}^{k} F_j^{[i]} \tag{8.105}
$$

We can notice that (8.103) and (8.104) require only one matrix inversion, which does not relate to the input signals, therefore the choice of the width of block pulses becomes more easy. But on the other hand, more additions and multiplications must be done to get satisfactory results from the iteration.

To illustrate the iterative algorithm, we list here some results of $X^{[k]}$ for Example 8.10:

$$X^{[1]}(t) \doteq \left(\begin{array}{ccccc} 0.1495 & 0.4470 & 0.7400 & 1.0258 & 1.3022 \\ 0.0882 & 0.2425 & 0.3555 & 0.4323 & 0.4776 \end{array} \right.$$

$$\left. \begin{array}{ccccc} 1.5675 & 1.8200 & 2.0586 & 2.2824 & 2.4905 \\ 0.4959 & 0.4909 & 0.4663 & 0.4253 & 0.3707 \end{array} \right) \Phi(t)$$

$$X^{[2]}(t) \doteq \left(\begin{array}{ccccc} 0.1613 & 0.5025 & 0.8754 & 1.2693 & 1.6749 \\ 0.0948 & 0.2724 & 0.4246 & 0.5490 & 0.6444 \end{array} \right.$$

$$\left. \begin{array}{ccccc} 2.0839 & 2.4895 & 2.8856 & 3.2676 & 3.6313 \\ 0.7109 & 0.7493 & 0.7610 & 0.7480 & 0.7124 \end{array} \right) \Phi(t)$$

$$X^{[3]}(t) \doteq \left(\begin{array}{ccccc} 0.1622 & 0.5083 & 0.8939 & 1.3112 & 1.7517 \\ 0.0953 & 0.2755 & 0.4344 & 0.5702 & 0.6815 \end{array} \right.$$

$$\left. \begin{array}{ccccc} 2.2070 & 2.6692 & 3.1307 & 3.5851 & 4.0263 \\ 0.7674 & 0.8273 & 0.8612 & 0.8697 & 0.8539 \end{array} \right) \Phi(t)$$

Comparing these vectors with (8.96), the convergence of the iterative results is obvious.

8.4.2 System identification

For both time-invariant and time-varying bilinear systems, our problem is to estimate the unknown parameter matrices from the block pulse coefficients of the inputs $U(t)$ and the states $X(t)$. Based on the relations between inputs and states in the block pulse domain, this identification problem can be solved easily.

Consider a time-invariant bilinear system (8.88). Based on (8.93), we can directly use the estimation algorithms (8.21) or (8.24) where only the matrices F, G, Θ or the vectors f_k, g_k should be reconstructed. For example, the matrices F, G and Θ are now:

$$F^T = \left(\begin{array}{cccc} X_2 - X_1 & X_3 - X_2 & \cdots & X_m - X_{m-1} \end{array} \right) \tag{8.106}$$

$$G^T = \frac{h}{2} \left(\begin{array}{cccc} X_2 + X_1 & X_3 + X_2 & \cdots & X_m + X_{m-1} \\ X_2 u_{1,2} + X_1 u_{1,1} & X_3 u_{1,3} + X_2 u_{1,2} & \cdots & X_m u_{1,m} + X_{m-1} u_{1,m-1} \\ \vdots & \vdots & \cdots & \vdots \\ X_2 u_{r,2} + X_1 u_{r,1} & X_3 u_{r,3} + X_2 u_{r,2} & \cdots & X_m u_{r,m} + X_{m-1} u_{r,m-1} \\ U_2 + U_1 & U_3 + U_2 & \cdots & U_m + U_{m-1} \end{array} \right) \tag{8.107}$$

and

$$\Theta^T = \left(\begin{array}{cccc} A & N_1 & \cdots & N_r & B \end{array} \right) \tag{8.108}$$

Example 8.11 Consider a bilinear system which is the same as in Example 8.10. Here the inputs are $u_1(t) = \sin(t) + \sin(2t)$ and $u_2(t) = \sin(3t) + \sin(4t)$. Our problem is to estimate the unknown parameter matrices from the simulated data of input and states. The sampling period is $h = 0.01$.

For this identification problem, we use the relation between block pulse coefficients of the inputs and states (8.93), and estimate the parameter matrices from the recursive least-squares algorithm (8.24). The estimation results are listed in Table 8.3.

From the above discussions, we notice that owing to the multiplication rule of block pulse series, bilinear systems can be manipulated as simply as linear systems in the block pulse domain. In fact, this operation rule of block pulse series benefits not only bilinear systems, but also a group of nonlinear systems whose terms contain products of state and input variables. For example, from a nonlinear system described by the state equation:

$$\dot{X}(t) = AX(t) + BU(t) + \sum_{i=1}^{n} C_i X(t)x_i(t) + \sum_{i=1}^{r} D_i X(t)u_i(t) + \sum_{i=1}^{r} E_i U(t)u_i(t) \quad (8.109)$$

we can directly obtain the relation between the block pulse coefficients of the input and

t	0.6	0.7	0.8	0.9	1.0	True value
$\hat{a}_{11}$	0.1364	0.1078	0.0898	0.0091	-0.0005	0.0
$\hat{a}_{12}$	0.8806	0.9056	0.9214	0.9920	1.0004	1.0
$\hat{a}_{21}$	-0.4428	-0.4554	-0.4622	-0.4925	-0.4978	-0.5
$\hat{a}_{22}$	-1.0499	-1.0390	-1.0330	-1.0065	-1.0019	-1.0
$\hat{b}_{11}$	0.4694	0.6945	0.9289	1.8827	1.9871	2.0
$\hat{b}_{12}$	1.3572	1.3144	1.2570	1.0317	1.0072	1.0
$\hat{b}_{21}$	0.3661	0.4652	0.5510	0.9100	0.9681	1.0
$\hat{b}_{22}$	1.1476	1.1286	1.1078	1.0230	1.0094	1.0
$\hat{n}_{1,11}$	-0.0502	-0.0418	-0.0338	0.0016	-0.0015	0.0
$\hat{n}_{1,12}$	1.0682	1.0666	1.0511	0.9959	0.9992	1.0
$\hat{n}_{1,21}$	0.4789	0.4826	0.4855	0.4977	0.4995	0.5
$\hat{n}_{1,22}$	0.2283	0.2275	0.2221	0.2004	0.2024	0.2
$\hat{n}_{2,11}$	1.0047	1.0051	1.0038	0.9996	0.9998	1.0
$\hat{n}_{2,12}$	-0.0210	-0.0213	-0.0161	0.0015	0.0005	0.0
$\hat{n}_{2,21}$	0.2019	0.2021	0.2017	0.2002	0.2000	0.2
$\hat{n}_{2,22}$	0.2913	0.2912	0.2929	0.2993	0.2999	0.3

Table 8.3: Recursive estimation results of a bilinear system.

output signals:

$$X_{k+1} - X_k = \frac{h}{2}A(X_{k+1} + X_k) + \frac{h}{2}B(U_{k+1} + U_k) + \frac{h}{2}\sum_{i=1}^{n} C_i(X_{k+1}x_{i,k+1} + X_k x_{i,k})$$

$$+ \frac{h}{2}\sum_{i=1}^{r} D_i(X_{k+1}u_{i,k+1} + X_k u_{i,k}) + \frac{h}{2}\sum_{i=1}^{r} E_i(U_{k+1}u_{i,k+1} + U_k u_{i,k}) \tag{8.110}$$

Discussions about block pulse function methods for solving analysis and identification problems of bilinear systems and their extensions can be found in the papers of Jan and Wong (1981), Cheng and Hsu (1982a), Wang (1982a,b), Ren, Shih, Pei and Guo (1983), Marszalek (1985a).

8.5 Hammerstein model nonlinear systems

A Hammerstein model nonlinear system is constructed by a linear subsystem and a memoryless nonlinear gain. In the state space representation, after introducing an intermediate variable $w(t)$ to relate the nonlinear gain:

$$w(t) = f(u(t)) \tag{8.111}$$

a time-invariant Hammerstein model nonlinear system can be described by:

$$\dot{X}(t) = AX(t) + Bw(t) \tag{8.112}$$

and a time-varying Hammerstein model nonlinear system can be described by:

$$\dot{X}(t) = A(t)X(t) + B(t)w(t) \tag{8.113}$$

where $X(t)$ is an n-dimensional state vector, and $u(t)$ is a scalar input. Since (8.112) and (8.113) have the same forms as the linear systems (8.1) and (8.47), the results discussed in Sections 8.1 and 8.2 can also be extended straightforward here.

8.5.1 System analysis

Our problem here is to evaluate the block pulse coefficients of the states $X(t)$ in a finite interval $t \in [0, T)$ from the state equations and the initial value of the states $X(0)$ under the input excitation $u(t)$. Since the analysis of linear systems and the manipulation of nonlinear gain have already been discussed in Section 8.1 and Section 7.5 respectively, the analysis problem of Hammerstein model nonlinear systems can be solved directly by combining the results of these two discussions. For example, for the time-invariant case (8.112), the combination of (7.91), (8.11) and (8.12) gives:

$$X_1 = \Gamma\left(X(0) + \frac{h}{2}B\sum_{i=1}^{p} r_i u_1^i\right) \tag{8.114}$$

and

$$X_{k+1} = \Gamma\left(\left(I + \frac{h}{2}A\right)X_k + \frac{h}{2}B\sum_{i=1}^{p} r_i\left(u_{k+1}^i + u_k^i\right)\right) \tag{8.115}$$

from which the block pulse coefficients of the states can be computed recursively.

8.5.2   System identification

For both time-invariant and time-varying Hammerstein model nonlinear systems, our problem is to estimate the unknown parameter matrices in the state equations and the coefficients of the polynomial of the nonlinear gain in (7.91) from the block pulse coefficients of the input $u(t)$ and the states $X(t)$.

Here we consider a time-invariant Hammerstein model nonlinear system (8.112). After combining (7.91) and (8.10):

$$X_{k+1} - X_k = \frac{h}{2} A \left(X_{k+1} + X_k \right) + \frac{h}{2} B \sum_{i=1}^{p} r_i \left(u_{k+1}^i + u_k^i \right) \tag{8.116}$$

and defining a set of vectors $(i = 1, 2, \ldots, p)$:

$$C_i = r_i B \tag{8.117}$$

the equation (8.116) can be rewritten as:

$$X_{k+1} - X_k = \frac{h}{2} A \left(X_{k+1} + X_k \right) + \frac{h}{2} \sum_{i=1}^{p} C_i \left(u_{k+1}^i + u_k^i \right) \tag{8.118}$$

Again, we can directly use the estimation algorithms (8.21) or (8.24), only the matrices F, G, Θ or the vectors f_k, g_k should be reconstructed. For example, the matrices F, G and Θ are now:

$$F^T = \left(\begin{array}{cccc} X_2 - X_1 & X_3 - X_2 & \cdots & X_m - X_{m-1} \end{array} \right) \tag{8.119}$$

$$G^T = \frac{h}{2} \left(\begin{array}{cccc} X_2 + X_1 & X_3 + X_2 & \cdots & X_m + X_{m-1} \\ u_2 + u_1 & u_3 + u_2 & \cdots & u_m + u_{m-1} \\ u_2^2 + u_1^2 & u_3^2 + u_2^2 & \cdots & u_m^p + u_{m-1}^p \\ \vdots & \vdots & \cdots & \vdots \\ u_2^p + u_1^p & u_3^p + u_2^p & \cdots & u_m^p + u_{m-1}^p \end{array} \right) \tag{8.120}$$

$$\Theta^T = \left(\begin{array}{ccccc} A & C_1 & C_2 & \ldots & C_p \end{array} \right) \tag{8.121}$$

In case of $r_1 = 1$, the parameter vector $\hat{B}$ and the coefficients in the power polynomial $\hat{r}_i$ $(i = 2, 3, \ldots, p)$ can be obtained after the matrix $\hat{A}$ and vectors $\hat{C}_i$ $(i = 1, 2, \ldots, p)$ are estimated:

$$\hat{B} = \hat{C}_1 \tag{8.122}$$

and

$$\hat{r}_i = \hat{B}^T \hat{C}_i / \hat{B}^T \hat{B} \tag{8.123}$$

Example 8.12 Consider a Hammerstein model nonlinear system described by (8.112) and (7.91). Here the parameter matrices are:

$$A = \left(\begin{array}{cc} 0 & -2 \\ 1 & -3 \end{array} \right), \quad B = \left(\begin{array}{c} 1 \\ 0 \end{array} \right)$$

and the nonlinear gain is:

$$w(t) = u(t) + 2u^2(t) + 3u^3(t) + 4u^4(t)$$

The input signal is $u(t) = \sin(t) + \sin(2t)$. Our problem is to estimate the unknown parameter matrices A and B together with the coefficients of the polynomial r_2, r_3 and r_4 from the simulated data of input and states. Here, the sampling period is $h = 0.05$.

For this identification problem, we use the relation between the block pulse coefficients of the input and the states in (8.116), and estimate the parameter matrices from the recursive least-squares algorithm (8.24). The estimation results are listed in Table 8.4.

Using block pulse functions, similar manipulations can also be applied to Hammerstein model nonlinear systems containing time delay. As an example, here is a time-invariant system:

$$\dot{X}(t) = AX(t) + Bf(u(t)) + LX(t - \tau) + Mf(u(t - \tau)) \tag{8.124}$$

Obviously, it is a combination of (8.71) and (8.112). If we also express the nonlinear gain $f(u(t))$ as a power polynomial (7.91) and the delay time as $\tau = qh$, from the expressions (8.82) and (8.115), we obtain:

$$X_1 = \Gamma\left(X(0) + \frac{h}{2}\left(B\sum_{i=1}^{p} r_i u_1^i + LX_{1-q} + M\sum_{i=1}^{p} r_i u_{1-q}^i\right)\right) \tag{8.125}$$

$$X_{k+1} = \Gamma\left(\left(I + \frac{h}{2}A\right)X_k + \frac{h}{2}B\sum_{i=1}^{p} r_i\left(u_{k+1}^i + u_k^i\right)\right.$$
$$\left. + \frac{h}{2}L(X_{k-q+1} + X_{k-q}) + \frac{h}{2}M\sum_{i=1}^{p} r_i\left(u_{k-q+1}^i + u_{k-q}^i\right)\right) \tag{8.126}$$

t	3.0	4.0	5.0	6.0	True value
$\hat{a}_{11}$	0.2612	0.0041	-0.0026	-0.0029	0.0
$\hat{a}_{12}$	-2.2571	-1.9988	-1.9919	-1.9916	-2.0
$\hat{a}_{21}$	0.3583	0.9880	1.0027	1.0026	1.0
$\hat{a}_{22}$	-2.3539	-2.9867	-3.0016	-3.0015	-3.0
$\hat{b}_1$	1.0402	1.0220	1.0204	1.0202	1.0
$\hat{b}_2$	-0.0503	-0.0061	-0.0026	-0.0027	0.0
$\hat{r}_2$	1.7855	1.9934	2.0056	2.0069	2.0
$\hat{r}_3$	3.1088	2.9497	2.9414	2.9407	3.0
$\hat{r}_4$	3.4891	3.8992	3.9173	3.9188	4.0

Table 8.4: Recursive estimation results of a Hammerstein model nonlinear system.

and

$$X_{k+1} - X_k = \frac{h}{2} A \left(X_{k+1} + X_k \right) + \frac{h}{2} B \sum_{i=1}^{p} r_i \left(u_{k+1}^i + u_k^i \right)$$

$$+ \frac{h}{2} L \left(X_{k-q+1} + X_{k-q} \right) + \frac{h}{2} M \sum_{i=1}^{p} r_i \left(u_{k-q+1}^i + u_{k-q}^i \right) \qquad (8.127)$$

Among them, (8.125) and (8.126) can be used to solve the system analysis problem, and (8.127) can be used to solve the system identification problem. Discussions about block pulse function methods for solving analysis and identification problems of Hammerstein model nonlinear systems can be found in the papers of Kung and Shih (1986), Chen and Lin (1986), Jiang (1987, 1988).

8.6   General nonlinear systems

Generally, a nonlinear system can be described by the state equation:

$$\dot{X}(t) = F(X(t), U(t), t) \qquad (8.128)$$

where $X(t)$ is an n-dimensional state vector, and $u(t)$ is a r-dimensional input vector. In using block pulse function methods, problems related to such nonlinear systems can be manipulated in a similar way as the various kinds of systems discussed before, i.e. the differential equations are first transformed into the equivalent integral equations, and the transformed integral equations are then approximated by their block pulse series. Thus, problems related to nonlinear systems can also be reduced to algebraic equations based on the operation rules of block pulse series.

8.6.1   System analysis

Integrating (8.128) from 0 to t on both sides, we have:

$$X(t) - X(0) = \int_0^t F(X(t), U(t), t) dt \qquad (8.129)$$

Expanding the functions $X(t)$, $U(t)$ and t into their block pulse series, we can obtain the block pulse series of the above equation:

$$\left(\begin{array}{cccc} X_1 & X_2 & \cdots & X_m \end{array} \right) \Phi(t) - \left(\begin{array}{cccc} X(0) & X(0) & \cdots & X(0) \end{array} \right) \Phi(t)$$

$$\doteq \left(\begin{array}{cccc} F(X_1, U_1, t_1) & F(X_2, U_2, t_2) & \cdots & F(X_m, U_m, t_m) \end{array} \right) P\Phi(t) \qquad (8.130)$$

Due to the special upper triangular form of the conventional integration operational matrix P, we can also obtain m equations of block pulse coefficients by equating the block pulse coefficients of $\phi_k(t)$ $(k = 1, 2, \ldots, m)$ on both sides of the above equation separately, e.g. the kth equation $E_{(k)}$ is:

$$X_k - X(0) = \frac{h}{2} F(X_k, U_k, t_k) + h \sum_{j=1}^{k-1} F(X_j, U_j, t_j) \qquad (8.131)$$

This equation can be used directly to solve the system analysis problem, i.e. to determine the block pulse coefficients of the state vector from the system equation (8.128) and the initial value of the states $X(0)$ under the input excitations $U(t)$ in a finite interval $t \in [0, T]$. Obviously, it is not so easy to use (8.131) because there are $n \times m$ simultaneous nonlinear algebraic equations. To simplify this problem, we apply the operation $E_{(k+1)} - E_{(k)}$ ($k = 1, 2, \ldots, m-1$) on the two neighbouring equations and obtain:

$$X_{k+1} - X_k = \frac{h}{2}F(X_{k+1}, U_{k+1}, t_{k+1}) + \frac{h}{2}F(X_k, U_k, t_k) \tag{8.132}$$

Thus, we can start from:

$$X_1 - X(0) = \frac{h}{2}F(X_1, U_1, t_1) \tag{8.133}$$

and then continue with (8.132) to determine the block pulse coefficients of the state vector successively. In such a way, only n nonlinear algebraic equations should be solved in each step, therefore the size of computation is reduced.

Example 8.13 Determine the block pulse coefficients of the state variables of a nonlinear system described by:

$$\begin{cases} \dot{x}_1 = \dfrac{a_1}{x_1} + a_2 x_2 + a_3 u(t) \\ \dot{x}_2 = \dfrac{a_4}{x_1} \end{cases}$$

in the interval $t \in [0, 1.2]$. Here, the parameters are $a_1 = a_2 = -1$ and $a_3 = a_4 = 1$, the initial values of the states are $x_1 = 1$ and $x_2 = 0$, and the input excitation is $u(t) = 1 + 0.5\sin(\pi t)$.

Set the number of block pulse functions to $m = 24$. According to (8.133) and (8.132), we can first obtain $x_{1,1} = 1.0004$ and $x_{2,1} = 0.0250$ from:

$$\begin{cases} x_{1,1} = 1 + \dfrac{h}{2}\left(u_1 - \dfrac{1}{x_{1,1}} - x_{2,1}\right) \\ x_{2,1} = \dfrac{h}{2x_{1,1}} \end{cases}$$

and then obtain $x_{1,2} = 1.0018$ and $x_{2,2} = 0.0749$ from:

$$\begin{cases} x_{1,2} - x_{1,1} = \dfrac{h}{2}\left(u_2 + u_1 - \dfrac{1}{x_{1,2}} - \dfrac{1}{x_{1,1}} - x_{2,2} - x_{2,1}\right) \\ x_{2,2} - x_{2,1} = \dfrac{h}{2}\left(\dfrac{1}{x_{1,2}} + \dfrac{1}{x_{1,1}}\right) \end{cases}$$

and so on. The piecewise constant approximations of the state variables and their exact solutions are illustrated in Figure 8.9 for comparison. In solving this problem via block pulse functions, we notice that nonlinear algebraic equations should be solved although we can directly transform the original differential equations to their corresponding algebraic expressions. In fact, it is not very efficient when block pulse functions are applied to the analysis problem of a general nonlinear system.

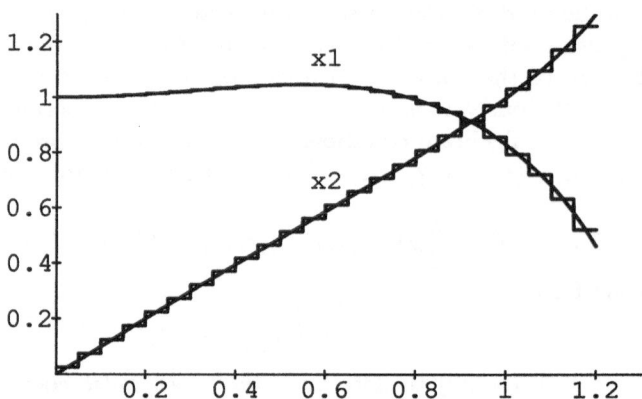

Figure 8.9: State variables of a nonlinear system.

8.6.2   System identification

To estimate the unknown parameters of nonlinear systems via block pulse functions, (8.132) is suitable to express the relations between the block pulse coefficients of the inputs and the states. Based on such relations, we can use the estimation algorithms (7.17), (7.18) or (8.24), (8.25) flexibly. Different from the block pulse function method in the system analysis case, no nonlinear algebraic equations are included here, therefore the block pulse function method for nonlinear system identification is much simpler.

Example 8.14 Consider a nonlinear system which is the same as in Example 8.13. Here the input is $u(t) = 1 + 0.5\sin(\pi t)$. Our problem is to estimate the unknown parameters a_i $(i = 1, 2, 3, 4)$ from the simulated data of the input and the states. The sampling period is $h = 0.05$.

Since the relation between block pulse coefficients of the inputs and the states is now:

$$x_{1,k+1} - x_{1,k} = a_1\frac{h}{2}\left(\frac{1}{x_{1,k+1}} + \frac{1}{x_{1,k}}\right) + a_2\frac{h}{2}\left(x_{2,k+1} + x_{2,k}\right) + a_3\frac{h}{2}(u_{k+1} + u_k) \quad (8.134)$$

and

$$x_{2,k+1} - x_{2,k} = a_4\frac{h}{2}\left(\frac{1}{x_{1,k+1}} + \frac{1}{x_{1,k}}\right) \quad (8.135)$$

we can estimate the first three parameters from (8.134) and estimate the last parameter from (8.135). Using the ordinary least-squares algorithm (7.17), the estimated results are $a_1 = -1.0004$, $a_2 = -0.9986$, $a_3 = 1.0005$ and $a_4 = 0.9997$, which are very close to the exact values of parameters. Discussions about block pulse function methods for solving analysis and identification problems of general nonlinear systems can be found in the papers of Shih (1978), Rao (1978), Rao and Srinivasan (1978a) Shieh and Yates (1979), Xing and Wang (1985).

Chapter 9

Practical aspects in using block pulse functions

For practical applications of the developed block pulse function methods, some problems should be discussed further. Since block pulse series of continuous signals are used in all these methods, the evaluation and approximation of block pulse coefficients from measured data should be considered. Since signals of real systems are always corrupted by noise, modification of identification algorithms for improving the biased estimation should also be considered. As practical examples, experiments are also demonstrated in this chapter to show the feasibility of the block pulse function methods.

9.1 Use of sampled data in block pulse function methods

As discussed before, the block pulse function technique is based on the block pulse series expansion. In order to use the operation rules of block pulse series, block pulse coefficients of continuous signals should be evaluated first. Strictly speaking, block pulse coefficients should be evaluated from (1.14), i.e.

$$f_k = \frac{1}{h} \int f(t)dt \Big|_{(k-1)h}^{kh} \qquad (9.1)$$

But in using this formula, m definite integrals are included which will complicate the solution of problems. Furthermore, such evaluation is possible only when the analytical expressions of continuous signals are known or when measurements between the sampling instants are available. For certain practical problems, such as system identification by means of digital computers, only the data of signals at sampling instants are available. In such cases, block pulse coefficients of the continuous signals must be approximated.

In the absence of any further information about the variations of signals between the sampling instants, the simplest approximation of block pulse coefficients is the mean value of the signal at the two end points of the corresponding subinterval, as formulated in (1.17):

$$f_k \doteq \frac{1}{2}(\bar{f}_{k-1} + \bar{f}_k) \tag{9.2}$$

where $\bar{f}_{k-1}$ and $\bar{f}_k$ are the sampled values of the continuous signal $f(t)$ at the time instants $t = (k-1)h$ and $t = kh$, respectively. Other formulas can also be used to improve the approximation of block pulse coefficients from the sampled data. As an example, here we use Stirling's interpolation formula to approximate block pulse coefficients. If $\hat{f}_{k-0.5}$ denotes the interpolation value of the function $f(t)$ at the middle point of the kth subinterval, the kth block pulse coefficient of $f(t)$ can be approximated by:

$$f_k \doteq \frac{1}{2}\left(\frac{\bar{f}_{k-1} + \hat{f}_{k-0.5}}{2} + \frac{\hat{f}_{k-0.5} + \bar{f}_k}{2}\right) \tag{9.3}$$

Using Stirling's interpolation formula with three points (Tuma, 1987):

$$\hat{y}_{k-0.5} \doteq \frac{1}{8}(3\bar{y}_k + 6\bar{y}_{k-1} - \bar{y}_{k-2}) \tag{9.4}$$

the kth block pulse coefficient can be approximated by:

$$f_k \doteq \frac{1}{16}(7\bar{f}_k + 10\bar{f}_{k-1} - \bar{f}_{k-2}) \tag{9.5}$$

Since (9.1) means that a block pulse coefficient is the mean height related to the area bounded by the continuous signal, (9.2) means that a block pulse coefficient is the mean height related to the area bounded by a trapezoid, and (9.3) means that a block pulse coefficient is the mean height related to the area bounded by two trapezoids in the corresponding subinterval, the block pulse coefficient obtained from the three points approximation (9.5) is usually better than the one obtained from the two points approximation (9.2). This improvement of block pulse coefficients is illustrated in Figure 9.1.

When problems are solved by block pulse function methods, the quality of the block pulse coefficients influences the results of simulation and identification problems in different ways. To show these different influences, some examples are given.

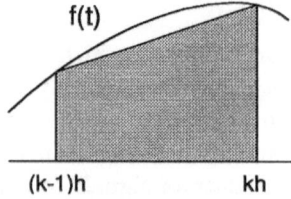
Two point interpolation

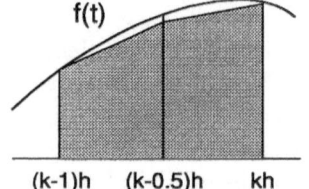
Three point interpolation

Figure 9.1: Different approximations of block pulse coefficients.

Example 9.1 Consider a second-order linear system described by (7.10):

$$G(s) = \frac{b_0}{s^2 + a_1 s + a_0}$$

with $a_0 = 2.0$, $a_1 = 3.0$ and $b_0 = 1.0$ under the input excitation $u(t) = \sin(t)$. Compare the simulation results under the block pulse difference equation by using the exact block pulse coefficients of the input obtained from (9.1) and their approximations obtained from (9.2) and (9.5).

A comparison procedure similar to the one described in Example 7.3 is used here. First, the block pulse coefficients of the input signal are evaluated from (9.1), (9.2) and (9.5) respectively, and then are applied to excite the block pulse difference equation (7.25). In these three cases, the obtained block pulse coefficients of the output signal are compared with the exact block pulse coefficients of the system response which are calculated directly from (7.27). The differences of the first 200 values are used to assess the quality of these three cases from the index of performance:

$$e_j = \sum_{k=1}^{200} |y_k - \tilde{y}_{j,k}| \qquad (9.6)$$

where y_k is the exact block pulse coefficient of the output, and $\tilde{y}_{j,k}$ $(j = 1, 2, 3)$ indicate the responses of the block pulse difference equation under the exact or approximate input block pulse coefficients respectively. The differences under various sampling periods are listed in Table 9.1. From these results, we notice that better approximation of block pulse coefficients of the input signal can provide better simulation results via block pulse function methods.

Example 9.2 Consider a second-order time-invariant linear system described by (7.10) with $a_0 = 2.0$, $a_1 = 3.0$ and $b_0 = 1.0$ under the continuous input excitation $u(t) = \sin(t) + \sin(2t)$. Estimate the unknown parameters from the sampled data of input and

Sampling period h	0.005	0.05	0.5
Method 1	0.000072	0.009883	0.778968
Method 2	0.000065	0.004404	0.4233414
Method 3	0.000063	0.003634	0.262874

Method 1 — Block pulse coefficients of input obtained from (9.2).
Method 2 — Block pulse coefficients of input obtained from (9.5).
Method 3 — Block pulse coefficients of input obtained from (9.1).

Table 9.1: Simulations under different block pulse coefficient approximations.

h	True value	Method 1	Method 2
	$a_0 = 2.0$	1.9998	1.9998
0.005	$a_1 = 3.0$	3.0003	3.0003
	$b_0 = 1.0$	1.0000	1.0000
	$a_0 = 2.0$	2.0015	2.0015
0.05	$a_1 = 3.0$	3.0034	3.0033
	$b_0 = 1.0$	1.0011	1.0011
	$a_0 = 2.0$	2.1300	2.1300
0.5	$a_1 = 3.0$	3.3272	3.3269
	$b_0 = 1.0$	1.1062	1.1059

Method 1 — Block pulse coefficients approximated by (9.2).
Method 2 — Block pulse coefficients approximated by (9.5).

Table 9.2: Estimations under different block pulse coefficient approximations.

output signals based on the block pulse regression equation. Compare the identification results under the different approximations of block pulse coefficients of the input and output signals which are obtained from (9.2) and (9.5).

In this comparison procedure, the data at 200 sampling instants are used. From these data, the block pulse coefficients of the input and output signals are evaluated by the two and three points approximation formulas respectively. Based on the block pulse regression equation (7.32), the ordinary least-squares algorithm (7.17) gives estimation results under different sampling intervals which are listed in Table 9.2. From these results, we notice that the estimated parameters can be improved when block pulse coefficients of the input and output signals are better approximated from their sampled data. But unlike in the simulation problem, the improvement of the estimation results due to better approximation of block pulse coefficients is hardly notable. Weighting the increase of computations and the improvement of estimations resulted by the better approximations of block pulse coefficients, we prefer to use the simplest formula (9.2) when identification problems are solved by the block pulse function methods. In fact, for improving the identification results via the block pulse function methods, a more efficient way is to reduce the sampling period. According to the rule proposed by Haykin (1972), for identifying all the modes of a continuous-time system using the bilinear z-transformation method, it is suitable to choose the sampling period h to satisfy $|p_k h| \leq 0.5$, where p_k is the system pole farthest from the origin of the s-plane. Since the estimation results based on the block pulse regression equation method are better than the bilinear z-transformation method (Jiang 1990), this rule is also feasible for choosing the sampling period in the identification problems using block pulse funtion methods.

9.2   Influence of noise in block pulse function methods

In the previous chapters, we used the least-squares algorithms to solve identification problems based on the block pulse function methods. The results of many examples show that the least-squares algorithms can provide satisfactory estimation results if no random noise is involved. In the discrete-time model identification, the biased estimation problem of ordinary least squares occurs, i.e. the estimations become asymptotically biased if the equation residuals are correlated. In fact, the same biased estimation phenomenon can also be seen in the identification methods based on block pulse functions, especially when the noise level is high. This is demonstrated clearly in the following examples.

Example 9.3 Consider the parameter estimation problem which is the same as in Example 9.2. But now a random white noise sequence $\{\bar{e}_k; k = 0, 1, \ldots\}$, which is normally distributed with mean 0.0 and variance 1.0, is added to the sampled output to simulate the noise corrupted output signal $\bar{y}_k$. The level of the noise is characterized by the noise-to-signal ratio:

$$N/S = \left(\frac{\sum \bar{e}_k^2}{\sum \bar{y}_k^2}\right)^{1/2} \tag{9.7}$$

If we use the block pulse regression equation here, (7.29) becomes:

$$z_{2,l} = -a_1 z_{1,l} - a_0 z_{0,l} + b_0 v_{0,l} \tag{9.8}$$

where the discrete values $z_{2,l}, z_{1,l}, z_{0,l}$ and $v_{0,l}$ are evaluated respectively from the block pulse coefficients y_{l+2}, y_{l+1}, y_l and u_{l+2}, u_{l+1}, u_l by relation (7.31). Setting the noise-to-signal ratio $N/S = 0.1$, the recursive least-squares algorithm yields the estimated parameters which are illustrated in Figure 9.2. Obviously, the estimations are strongly biased.

In discrete-time model identification, this biased estimation problem can be solved by various techniques, e.g. the instrumental variable method (Söderström and Stoica 1983). Certain variants of the instrumental variable method can also be applied to block pulse regression equation method for the continuous-time model identification. As a simple variant, the instrumental variables can be constructed directly from the data of the input signal (Wouters 1972). This construction of instrumental variables can be directly realized in the method of using block pulse regression equations. In the lth step estimation, the instrumental variables $\hat{z}_{2,l}, \hat{z}_{1,l}, \hat{z}_{0,l}$ and $\hat{v}_{0,l}$ can be evaluated respectively from the block pulse coefficients u_{l+2}, u_{l+1}, u_l and $u_{l-1}, u_{l-2}, u_{l-3}$ according to (7.31). As illustrated in Figure 9.3, this simple variant of instrumental variables improves the biased parameter estimation.

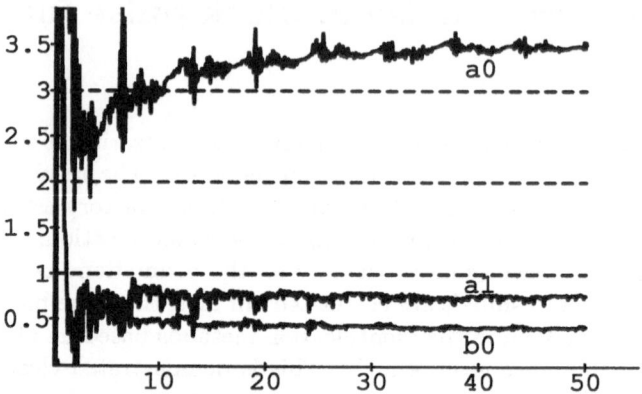

Figure 9.2: Least-squares algorithm yields biased estimation results.

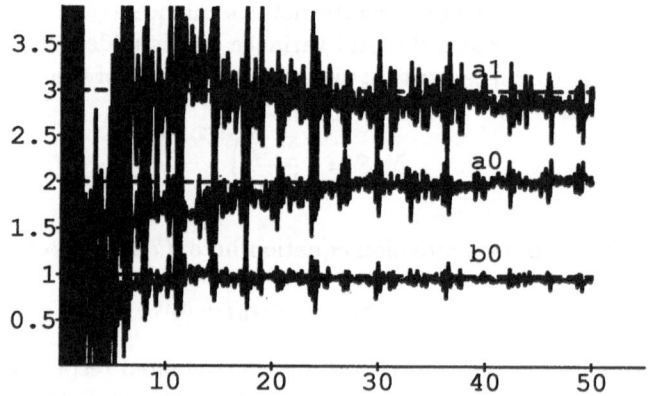

Figure 9.3: Instrumental variable algorithm improves the estimation results.

In the parameter estimation via block pulse function methods, noise levels and data lengthes also influence the results. To show these influences, we consider the same identification problem as in Example 9.3. Here, we set the sampling period $h = 0.05$ and use both the least-squares and instrumental variable algorithms described above. The batch processing estimation results related to data at 200 sampling instants under various levels of noise are listed in Table 9.3. The batch processing estimation results related to different data lengths under the noise level $N/S = 0.2$ are listed in Table 9.4. Similar to Figures 9.2 and 9.3, these estimation results also show that the least-squares algorithm is not a suitable scheme for identification problems using block pulse function methods when the noise level is high, and that the instrumental variable algorithms can overcome the biased estimation problem.

As another method to improve the biased estimation in the identification using block pulse function methods, we use here both the simulation and optimization techniques. At the beginning of the estimation procedure, we assign some values to the unknown

	N/S	$\hat{a}_0$	$\hat{a}_1$	$\hat{b}_0$
LS method	0.01	2.0043	2.9284	0.9820
	0.02	2.0122	2.7060	0.9263
	0.05	2.1557	1.7770	0.6993
	0.1	3.0152	0.8604	0.5019
	0.2	7.0317	0.3846	0.5304
IV method	0.01	2.0032	3.0131	1.0030
	0.02	2.0042	3.0216	1.0044
	0.05	2.0083	3.0533	1.0100
	0.1	2.0197	3.1306	1.0255
	0.2	2.0683	3.4215	1.0900
True value		2.0	3.0	1.0

Table 9.3: Influence of noise in the estimation results under different noise levels.

	m	$\hat{a}_0$	$\hat{a}_1$	$\hat{b}_0$
LS method	200	7.0317	0.3846	0.5304
	500	7.3930	0.3262	0.4374
	1000	7.4307	0.2898	0.3276
	2000	7.4312	0.2645	0.2797
	5000	7.4784	0.2443	0.2600
IV method	200	2.0683	3.4215	1.0900
	500	2.0090	3.1111	1.0259
	1000	2.0407	3.1053	1.0177
	2000	1.9928	2.9667	0.9718
	5000	1.9939	2.9583	0.9829
True value		2.0	3.0	1.0

Table 9.4: Influence of noise in the estimation results under different data lengthes.

system parameters. Since these values of parameters can only roughly be determined from the incomplete a priori knowledge, they are certainly incorrect. But based on them, simulation of the continuous system can be done through the corresponding block pulse difference equation. From (7.22) and the input signal, the block pulse coefficients of the output signal are first evaluated, and then a loss function like:

$$J = \sum_{k=1}^{m}(y_k - \hat{y}_k)^2 \qquad (9.9)$$

is used to assess the quality of the values of parameters, where y_k and $\hat{y}_k$ $(k = 1, 2, \ldots, m)$ are the block pulse coefficients of the output signal obtained from the measured and the simulated data respectively. Through minimizing the loss function, we get the improved values of the parameters, and based on these new values, simulation can be done once again. In this estimation procedure with alternating simulation and optimization, the influence of noise can be decreased and the biased estimation of the ordinary least-squares algorithms can be improved because the optimization process forces the simulated output to coincide to the measured output in the whole.

Example 9.4 Consider a third-order time-invariant linear system described by the transfer function:

$$G(s) = \frac{b_0}{s^3 + a_2 s^2 + a_1 s + a_0}$$

with unknown parameters. From the continuous input excitation, the system output is first generated and corrupted by random noise with the noise-to-signal ratio $N/S = 0.2$. Our task is to estimate the unknown parameters from the sampled data of the input and output signals using the above described simulation and optimization procedure. The sampling period here is $h = 1$.

Starting from the roughly determined values of the unknown system parameters $a_0 = 6.0$, $a_1 = 11.0$, $a_2 = 6$ and $b_0 = 1$, we first obtain the simulated output from the block pulse difference equation (7.22), which is illustrated in Figure 9.4 together with the noisy sampled data of the system output. Obviously, the strong incoincidence of the two output signals indicates the wrong values of parameters before the optimization.

To improve the estimation results, the Powell algorithm is used in the optimization process. Based on the improved values of the system parameters $a_0 = 17.5$, $a_1 = 18.6$,

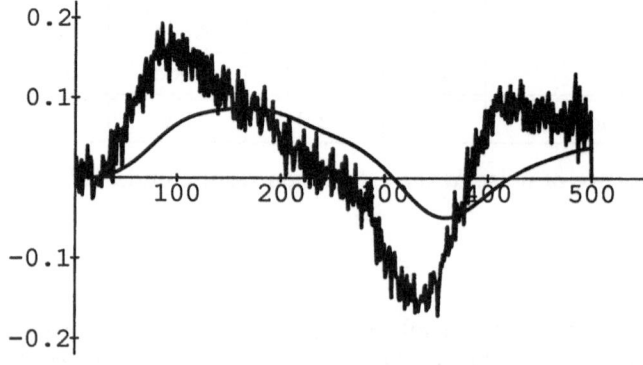

Figure 9.4: Output signals before optimization.

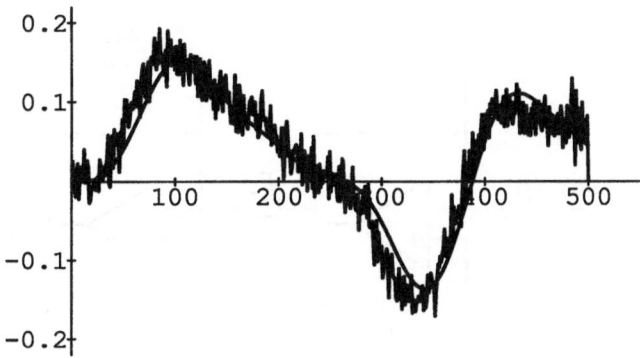

Figure 9.5: Output signals after optimization.

$a_2 = 4.5$ and $b_0 = 2.7$, the simulated output together with the noisy sampled data of the output signal are illustrated in Figure 9.5. Obviously, the two output signals coincide with each other after the optimization although the noise level is high. This example shows that the iterative identification approach based on the simulation and optimization techniques can easily be applied to decrease the influence of noise when block pulse difference equations are used in the simulation stage of continuous-time systems.

9.3 Practical applications of block pulse function methods

To demonstrate practical applications of block pulse function methods, we describe here two experiments which are related to system identification using the block pulse regression equations.

The first experiment is to estimate the parameters of the transfer function of an electrical circuit, which is composed of operational amplifiers, resistances and capacitors, as illustrated in Figure 9.6. In the circuit, the resistances are measured as $R_{11} = 997.8\Omega$, $R_{12} = 997.0\Omega$, $R_{21} = 999.2\Omega$, $R_{22} = 1000.0\Omega$ and the capacitances are $C_1 = 35.85\mu F$, $C_2 = 47.95\mu F$, respectively. From the construction of the circuit, a second-order system can be obtained which is described by the transfer function (7.10) with $a_0 = 583.48$, $a_1 = 48.83$ and $b_0 = 583.50$. Now, without measuring the values of the resistances and capacitors, the task of the experiment is to estimate the unknown parameters a_0, a_1 and b_0 from the measured data of the input and output signals.

The system input is composed of two sine signals, which are produced by signal generators. They have the frequencies 2Hz, 3Hz and the amplitudes 3.0V, 3.5V, respectively. A SICOMP PC 16-05 personal computer, in which a Burr-Brown PCI-20000 system is installed, is used to fulfill the tasks of data acquisition and parameter estima-

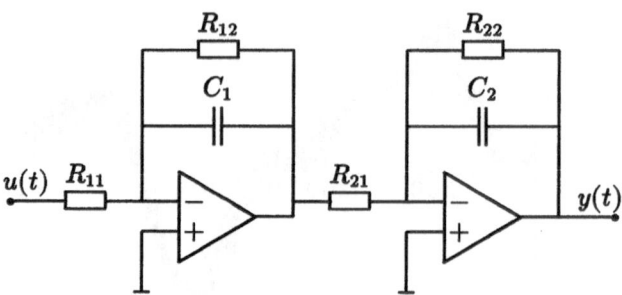

Figure 9.6: The experimental electrical circuit.

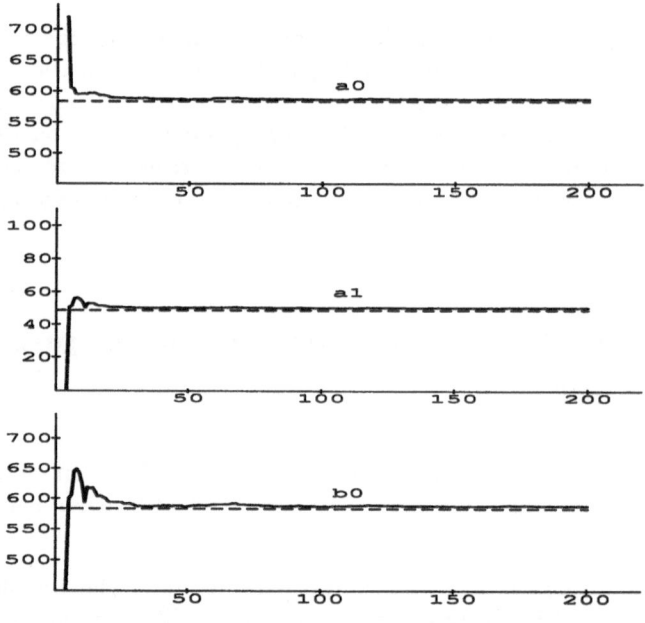

——— : estimated parameters,   - - - : true parameters

Figure 9.7: Estimated parameters in the experiment of electrical circuit.

tion. For the A/D convertors, the full range of the analog signal is ±10V. The resolution of these convertors is 12 bits, and it determines the accuracy of the sampled data of the signals. Except for the quantization error in the data acquisition, no extra noise is introduced in this system. The sampling rate is about 75Hz.

Since the noise level is low in this experiment, the recursive least-squares algorithm based on the block pulse regression equation (9.8) is used for the parameter estimation.

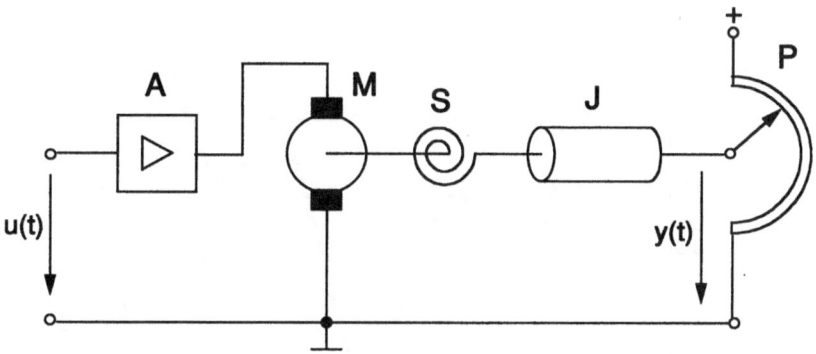

Figure 9.8: The experimental DC-servo system.

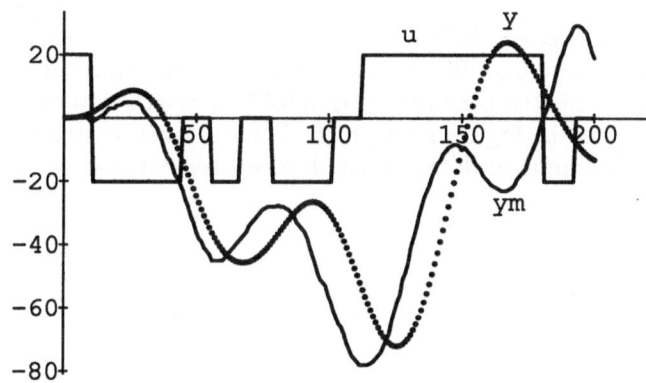

Figure 9.9: Input and output signals before optimization.

For each step of recursion, the time of about 12ms is needed for data acquisition and computation. The estimated parameters converge quickly as illustrated in Figure 9.7. After about 20 recursions, the estimated parameters converge to the values $\hat{a}_0 = 586$, $\hat{a}_1 = 50$ and $\hat{b}_0 = 587$, with relative errors of about 0.4%, 2.4% and 0.6%, respectively.

The second experiment is to identify a DC-servo system from the measurements of the input signal $u(t)$ and the output signal $y(t)$. The system is shown in Figure 9.8, where the symbols A, M, S, J and P stand for the amplifier, the electric motor, the spring, the mass and the potentiometer, respectively. From physical considerations, the basic model of the system can be expressed by the transfer function:

$$G(s) = \frac{b_0}{s^3 + a_2 s^2 + a_1 s} \tag{9.10}$$

The unknown parameters a_1, a_2 and b_0 can be estimated by the method of combining both simulation and optimization as described in the previous section. Starting from the incorrect values of parameters $\hat{a}_2 = 0.35$, $\hat{a}_1 = 22$ and $\hat{b}_0 = 154$, the simulated

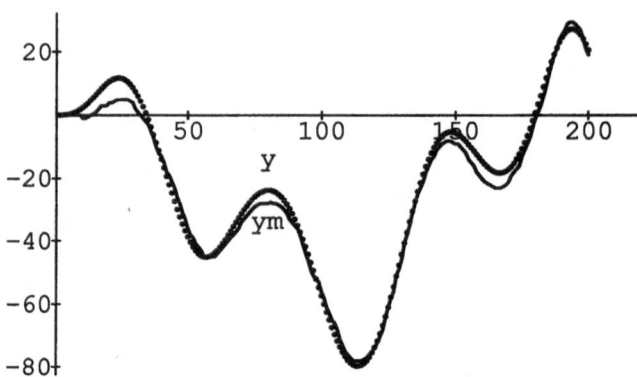

Figure 9.10: Output signals after optimization.

output signal y is obtained. It is illustrated together with the real input signal u and the output signal y_m ($h=5.0/200$sec.) in Figure 9.9. The optimization process gives the final result of the estimated parameters $a_2 = 0.57$, $a_1 = 34.5$ and $b_0 = 269$. Both output signals are illustrated in Figure 9.10 which shows a much better coincidence between the measured and simulated data, although measurement noise and nonlinearity exist in this system.

Appendix A

Some block pulse integration operational matrices

With $m = 7$, the block pulse integration operational matrices $P_{i,j}$ $(i, j = 0, 1, 2, 3)$ are listed here for showing their regular constructions. These matrices are related to the operation rule of the block pulse series:

$$\underbrace{\int_0^t \cdots \int_0^t t^j f(t) \, dt \cdots dt}_{i \text{ times}} \doteq F^T P_{i,j} \Phi(t)$$

as discussed in Sections 5.5 and 5.6.

$$P_{0,0} = \begin{pmatrix} 1 & 0 & 0 & 0 & 0 & 0 & 0 \\ 0 & 1 & 0 & 0 & 0 & 0 & 0 \\ 0 & 0 & 1 & 0 & 0 & 0 & 0 \\ 0 & 0 & 0 & 1 & 0 & 0 & 0 \\ 0 & 0 & 0 & 0 & 1 & 0 & 0 \\ 0 & 0 & 0 & 0 & 0 & 1 & 0 \\ 0 & 0 & 0 & 0 & 0 & 0 & 1 \end{pmatrix}$$

$$P_{1,0} = \frac{h}{2} \begin{pmatrix} 1 & 2 & 2 & 2 & 2 & 2 & 2 \\ 0 & 1 & 2 & 2 & 2 & 2 & 2 \\ 0 & 0 & 1 & 2 & 2 & 2 & 2 \\ 0 & 0 & 0 & 1 & 2 & 2 & 2 \\ 0 & 0 & 0 & 0 & 1 & 2 & 2 \\ 0 & 0 & 0 & 0 & 0 & 1 & 2 \\ 0 & 0 & 0 & 0 & 0 & 0 & 1 \end{pmatrix}$$

$$P_{2,0} = \frac{h^2}{6} \begin{pmatrix} 1 & 6 & 12 & 18 & 24 & 30 & 36 \\ 0 & 1 & 6 & 12 & 18 & 24 & 30 \\ 0 & 0 & 1 & 6 & 12 & 18 & 24 \\ 0 & 0 & 0 & 1 & 6 & 12 & 18 \\ 0 & 0 & 0 & 0 & 1 & 6 & 12 \\ 0 & 0 & 0 & 0 & 0 & 1 & 6 \\ 0 & 0 & 0 & 0 & 0 & 0 & 1 \end{pmatrix}$$

$$P_{3,0} = \frac{h^3}{24} \begin{pmatrix} 1 & 14 & 50 & 110 & 194 & 302 & 434 \\ 0 & 1 & 14 & 50 & 110 & 194 & 302 \\ 0 & 0 & 1 & 14 & 50 & 110 & 194 \\ 0 & 0 & 0 & 1 & 14 & 50 & 110 \\ 0 & 0 & 0 & 0 & 1 & 14 & 50 \\ 0 & 0 & 0 & 0 & 0 & 1 & 14 \\ 0 & 0 & 0 & 0 & 0 & 0 & 1 \end{pmatrix}$$

$$P_{0,1} = \frac{h}{2} \begin{pmatrix} 1 & 0 & 0 & 0 & 0 & 0 & 0 \\ 0 & 3 & 0 & 0 & 0 & 0 & 0 \\ 0 & 0 & 5 & 0 & 0 & 0 & 0 \\ 0 & 0 & 0 & 7 & 0 & 0 & 0 \\ 0 & 0 & 0 & 0 & 9 & 0 & 0 \\ 0 & 0 & 0 & 0 & 0 & 11 & 0 \\ 0 & 0 & 0 & 0 & 0 & 0 & 13 \end{pmatrix}$$

$$P_{1,1} = \frac{h^2}{6} \begin{pmatrix} 1 & 3 & 3 & 3 & 3 & 3 & 3 \\ 0 & 4 & 9 & 9 & 9 & 9 & 9 \\ 0 & 0 & 7 & 15 & 15 & 15 & 15 \\ 0 & 0 & 0 & 10 & 21 & 21 & 21 \\ 0 & 0 & 0 & 0 & 13 & 27 & 27 \\ 0 & 0 & 0 & 0 & 0 & 16 & 33 \\ 0 & 0 & 0 & 0 & 0 & 0 & 19 \end{pmatrix}$$

$$P_{2,1} = \frac{h^3}{24} \begin{pmatrix} 1 & 10 & 22 & 34 & 46 & 58 & 70 \\ 0 & 5 & 34 & 70 & 106 & 142 & 178 \\ 0 & 0 & 9 & 58 & 118 & 178 & 238 \\ 0 & 0 & 0 & 13 & 82 & 166 & 250 \\ 0 & 0 & 0 & 0 & 17 & 106 & 214 \\ 0 & 0 & 0 & 0 & 0 & 21 & 130 \\ 0 & 0 & 0 & 0 & 0 & 0 & 25 \end{pmatrix}$$

$$P_{3,1} = \frac{h^4}{120} \begin{pmatrix} 1 & 25 & 105 & 245 & 445 & 705 & 1025 \\ 0 & 6 & 95 & 355 & 795 & 1415 & 2215 \\ 0 & 0 & 11 & 165 & 605 & 1345 & 2385 \\ 0 & 0 & 0 & 16 & 235 & 855 & 1895 \\ 0 & 0 & 0 & 0 & 21 & 305 & 1105 \\ 0 & 0 & 0 & 0 & 0 & 26 & 375 \\ 0 & 0 & 0 & 0 & 0 & 0 & 31 \end{pmatrix}$$

$$P_{0,2} = \frac{h^2}{3} \begin{pmatrix} 1 & 0 & 0 & 0 & 0 & 0 & 0 \\ 0 & 7 & 0 & 0 & 0 & 0 & 0 \\ 0 & 0 & 19 & 0 & 0 & 0 & 0 \\ 0 & 0 & 0 & 37 & 0 & 0 & 0 \\ 0 & 0 & 0 & 0 & 61 & 0 & 0 \\ 0 & 0 & 0 & 0 & 0 & 91 & 0 \\ 0 & 0 & 0 & 0 & 0 & 0 & 127 \end{pmatrix}$$

$$P_{1,2} = \frac{h^3}{12} \begin{pmatrix} 1 & 4 & 4 & 4 & 4 & 4 & 4 \\ 0 & 11 & 28 & 28 & 28 & 28 & 28 \\ 0 & 0 & 33 & 76 & 76 & 76 & 76 \\ 0 & 0 & 0 & 67 & 148 & 148 & 148 \\ 0 & 0 & 0 & 0 & 113 & 244 & 244 \\ 0 & 0 & 0 & 0 & 0 & 171 & 364 \\ 0 & 0 & 0 & 0 & 0 & 0 & 241 \end{pmatrix}$$

$$P_{2,2} = \frac{h^4}{60} \begin{pmatrix} 1 & 15 & 35 & 55 & 75 & 95 & 115 \\ 0 & 16 & 125 & 265 & 405 & 545 & 685 \\ 0 & 0 & 51 & 355 & 735 & 1115 & 1495 \\ 0 & 0 & 0 & 106 & 705 & 1445 & 2185 \\ 0 & 0 & 0 & 0 & 181 & 1175 & 2395 \\ 0 & 0 & 0 & 0 & 0 & 276 & 1765 \\ 0 & 0 & 0 & 0 & 0 & 0 & 391 \end{pmatrix}$$

$$P_{3,2} = \frac{h^5}{360} \begin{pmatrix} 1 & 41 & 191 & 461 & 851 & 1361 & 1991 \\ 0 & 22 & 401 & 1571 & 3581 & 6431 & 10121 \\ 0 & 0 & 73 & 1181 & 4451 & 10001 & 17831 \\ 0 & 0 & 0 & 154 & 2381 & 8831 & 19721 \\ 0 & 0 & 0 & 0 & 265 & 4001 & 14711 \\ 0 & 0 & 0 & 0 & 0 & 406 & 6041 \\ 0 & 0 & 0 & 0 & 0 & 0 & 577 \end{pmatrix}$$

$$P_{0,3} = \frac{h^3}{4} \begin{pmatrix} 1 & 0 & 0 & 0 & 0 & 0 & 0 \\ 0 & 15 & 0 & 0 & 0 & 0 & 0 \\ 0 & 0 & 65 & 0 & 0 & 0 & 0 \\ 0 & 0 & 0 & 175 & 0 & 0 & 0 \\ 0 & 0 & 0 & 0 & 369 & 0 & 0 \\ 0 & 0 & 0 & 0 & 0 & 671 & 0 \\ 0 & 0 & 0 & 0 & 0 & 0 & 1105 \end{pmatrix}$$

$$P_{1,3} = \frac{h^4}{20} \begin{pmatrix} 1 & 5 & 5 & 5 & 5 & 5 & 5 \\ 0 & 26 & 75 & 75 & 75 & 75 & 75 \\ 0 & 0 & 131 & 325 & 325 & 325 & 325 \\ 0 & 0 & 0 & 376 & 875 & 875 & 875 \\ 0 & 0 & 0 & 0 & 821 & 1845 & 1845 \\ 0 & 0 & 0 & 0 & 0 & 1526 & 3355 \\ 0 & 0 & 0 & 0 & 0 & 0 & 2551 \end{pmatrix}$$

$$P_{2,3} = \frac{h^5}{120} \begin{pmatrix} 1 & 21 & 51 & 81 & 111 & 141 & 171 \\ 0 & 42 & 381 & 831 & 1281 & 1731 & 2181 \\ 0 & 0 & 233 & 1761 & 3711 & 5661 & 7611 \\ 0 & 0 & 0 & 694 & 4881 & 10131 & 15381 \\ 0 & 0 & 0 & 0 & 1545 & 10461 & 21531 \\ 0 & 0 & 0 & 0 & 0 & 2906 & 19221 \\ 0 & 0 & 0 & 0 & 0 & 0 & 4897 \end{pmatrix}$$

$$P_{3,3} = \frac{h^6}{840} \begin{pmatrix} 1 & 63 & 315 & 777 & 1449 & 2331 & 3423 \\ 0 & 64 & 1365 & 5607 & 12999 & 23541 & 37233 \\ 0 & 0 & 379 & 6657 & 25809 & 58611 & 105063 \\ 0 & 0 & 0 & 1156 & 18879 & 71421 & 160713 \\ 0 & 0 & 0 & 0 & 2605 & 40971 & 152943 \\ 0 & 0 & 0 & 0 & 0 & 4936 & 75873 \\ 0 & 0 & 0 & 0 & 0 & 0 & 8359 \end{pmatrix}$$

Appendix B

Some formulas about upper triangular matrices

In the discussions about the block pulse functions, upper triangular matrices appear frequently. In order to save computations in certain operations with respect to the upper triangular matrices, equivalent formulas of the summation forms can be used for each entry in the results. Some of them are listed in this appendix for the convenience of use. They are divided into two groups, i.e. the formulas in the first group are concerned with the general upper triangular matrices which have the form:

$$
A = \begin{pmatrix}
a_{1,1} & a_{1,2} & \cdots & a_{1,m} \\
0 & a_{2,2} & \cdots & a_{2,m} \\
\vdots & \vdots & \ddots & \vdots \\
0 & 0 & \cdots & a_{m,m}
\end{pmatrix}
\tag{B.1}
$$

and the formulas in the second group are concerned with the special upper triangular matrices which have the form:

$$
A = \begin{pmatrix}
a_1 & a_2 & \cdots & a_m \\
0 & a_1 & \cdots & a_{m-1} \\
\vdots & \vdots & \ddots & \vdots \\
0 & 0 & \cdots & a_1
\end{pmatrix}
\tag{B.2}
$$

In both groups, X, Y denote the column vectors.

Formulas about general upper triangular matrices

Suppose that A and B are m-dimensional upper triangular matrices of the general form (B.1).

1. Product of a vector and a matrix. The entries of the vector $Y^T = X^T A$ can be evaluated from:

$$y_i = \sum_{k=1}^{i} x_k a_{k,i} \tag{B.3}$$

where $i = 1, 2, \ldots, m$. Equation (B.3) is merely the usual multiplication rule of a vector and a matrix, only the computation is reduced in the sum due to the zero lower triangular part of the multiplicator matrix.

2. Product of two matrices. The entries of the matrix $C = AB$ can be evaluated from:

$$c_{i,j} = \begin{cases} 0 & \text{for } j < i \\ \displaystyle\sum_{k=i}^{j} a_{i,k} b_{k,j} & \text{for } j \geq i \end{cases} \tag{B.4}$$

where $i, j = 1, 2, \ldots, m$. Equation (B.4) is merely the usual multiplication rule of two matrices, only the computation is reduced in the sum due to the zero lower triangular parts of both the multiplicand and multiplicator matrices.

3. Inverse of a matrix. The entries of the matrix $C = A^{-1}$ can be evaluated from:

$$c_{i,j} = \begin{cases} 0 & \text{for } j < i \\ \dfrac{1}{a_{i,i}} & \text{for } j = i \\ -\dfrac{1}{a_{i,i}} \displaystyle\sum_{k=i+1}^{j} a_{i,k} c_{k,j} & \text{for } j > i \end{cases} \tag{B.5}$$

where $i, j = 1, 2, \ldots, m$. Equation (B.5) can be obtained from (B.4), because the product of C and A is a unit matrix.

4. Equation with an unknown multiplicand matrix. The entries of the unknown multiplicand matrix C in the equation $A = CB$ can be evaluated from:

$$c_{i,j} = \begin{cases} 0 & \text{for } j < i \\ \dfrac{1}{b_{j,j}} a_{i,j} & \text{for } j = i \\ \dfrac{1}{b_{j,j}} \left(a_{i,j} - \displaystyle\sum_{k=i}^{j-1} c_{i,k} b_{k,j} \right) & \text{for } j > i \end{cases} \tag{B.6}$$

where $i, j = 1, 2, \ldots, m$. Equation (B.6) can be obtained from (B.4).

5. Equation with an unknown multiplicator matrix. The entries of the unknown multiplicator matrix C in the equation $A = BC$ can be evaluated from:

$$
c_{i,j} = \begin{cases}
0 & \text{for } j < i \\
\dfrac{1}{b_{i,i}} a_{i,j} & \text{for } j = i \\
\dfrac{1}{b_{i,i}} \left(a_{i,j} - \displaystyle\sum_{k=i+1}^{j} b_{i,k} c_{k,j} \right) & \text{for } j > i
\end{cases}
\tag{B.7}
$$

where $i, j = 1, 2, \ldots, m$. Equation (B.7) can be obtained from (B.4).

In (B.4), (B.5), (B.6) and (B.7), all the result matrices C are also upper triangular matrices of the general form (B.1).

Formulas about special upper triangular matrices

Suppose that A and B are m-dimensional upper triangular matrices of the special form (B.2). Since the entry $a_{i,j}$ $(i, j = 1, 2, \ldots, m)$ of the matrix A in (B.1) is now the entry a_{j-i+1} of the matrix A in (B.2), it is easy to derive the formulas about the special upper triangular matrices from the above results of the general cases.

1. Product of a vector and a matrix. The entries of the vector $Y^T = X^T A$ can be evaluated from:

$$
y_i = \sum_{k=1}^{i} x_k a_{i-k+1}
\tag{B.8}
$$

where $i = 1, 2, \ldots, m$.

2. Product of two matrices. The entries of the matrix $C = AB$ can be evaluated from:

$$
c_i = \sum_{k=1}^{i} a_k b_{i-k+1}
\tag{B.9}
$$

where $i = 1, 2, \ldots, m$. Equation (B.9) indicates that the multiplication of the special upper triangular matrices is interchangeable.

3. Inverse of a matrix. The entries of the matrix $C = A^{-1}$ can be evaluated from:

$$
c_i = \begin{cases}
\dfrac{1}{a_1} & \text{for } i = 1 \\
-\dfrac{1}{a_1} \displaystyle\sum_{k=2}^{i} a_k c_{i-k+1} & \text{for } i = 2, 3, \ldots, m
\end{cases}
\tag{B.10}
$$

4. Equation with an unknown matrix. The entries of the unknown matrix C in the equation $A = CB$ or in the equation $A = BC$ can be evaluated from:

$$
c_i = \begin{cases} \dfrac{1}{b_1} a_1 & \text{for } i = 1 \\[3ex] \dfrac{1}{b_1} \left(a_i - \displaystyle\sum_{k=1}^{i-1} c_k b_{i-k+1} \right) & \text{for } i = 2, 3, \ldots, m \end{cases} \tag{B.11}
$$

Here, we do not distinguish whether the unknown matrix in the equation is multiplicand or multiplicator, because the multiplication of the special upper triangular matrices is interchangeable.

5. Equation with an unknown matrix. The entries of the unknown matrix C in the equation $Y^T = X^T C$ can be evaluated from:

$$
c_i = \begin{cases} \dfrac{1}{x_1} y_1 & \text{for } i = 1 \\[3ex] \dfrac{1}{x_1} \left(y_i - \displaystyle\sum_{k=2}^{i} x_k c_{i-k+1} \right) & \text{for } i = 2, 3, \ldots, m \end{cases} \tag{B.12}
$$

Equation (B.12) can be obtained from (B.8).

In (B.9), (B.10), (B.11) and (B.12), all the result matrices C are also upper triangular matrices of the special form (B.1).

Appendix C

Some relations in block pulse difference equations

Some relations between the block pulse difference equations and their original differential equations are listed here for the convenient use.

1. Relations between parameters in (7.1) and (7.22). The following equations give the relations between the parameters a_i $(i = 0, 1, \ldots, n)$ in the differential equation (7.1) and the parameters A_j $(j = 0, 1, \ldots, n)$ in the block pulse difference equation (7.22). Similar relations also exist between the parameters b_i and B_j.

The 1st-order system:

$$A_1 = a_1 + \frac{1}{2}ha_0$$

$$A_0 = -a_1 + \frac{1}{2}ha_0$$

The 2nd-order system:

$$A_2 = a_2 + \frac{1}{2}ha_1 + \frac{1}{6}h^2a_0$$

$$A_1 = -2a_2 + \frac{2}{3}h^2a_0$$

$$A_0 = a_2 - \frac{1}{2}ha_1 + \frac{1}{6}h^2a_0$$

The 3rd-order system:

$$A_3 = a_3 + \frac{1}{2}ha_2 + \frac{1}{6}h^2a_1 + \frac{1}{24}h^3a_0$$

$$A_2 = -3a_3 - \frac{1}{2}ha_2 + \frac{1}{2}h^2a_1 + \frac{11}{24}h^3a_0$$

$$A_1 = 3a_3 - \frac{1}{2}ha_2 - \frac{1}{2}h^2a_1 + \frac{11}{24}h^3a_0$$

$$A_0 = -a_3 + \frac{1}{2}ha_2 - \frac{1}{6}h^2a_1 + \frac{1}{24}h^3a_0$$

The 4th-order system:

$$A_4 = a_4 + \frac{1}{2}ha_3 + \frac{1}{6}h^2a_2 + \frac{1}{24}h^3a_1 + \frac{1}{120}h^4a_0$$

$$A_3 = -4a_4 - ha_3 + \frac{1}{3}h^2a_2 + \frac{5}{12}h^3a_1 + \frac{13}{60}h^4a_0$$

$$A_2 = 6a_4 - h^2a_2 + \frac{11}{20}h^4a_0$$

$$A_1 = -4a_4 + ha_3 + \frac{1}{3}h^2a_2 - \frac{5}{12}h^3a_1 + \frac{13}{60}h^4a_0$$

$$A_0 = a_4 - \frac{1}{2}ha_3 + \frac{1}{6}h^2a_2 - \frac{1}{24}h^3a_1 + \frac{1}{120}h^4a_0$$

The 5th-order system:

$$A_5 = a_5 + \frac{1}{2}ha_4 + \frac{1}{6}h^2a_3 + \frac{1}{24}h^3a_2 + \frac{1}{120}h^4a_1 + \frac{1}{720}h^5a_0$$

$$A_4 = -5a_5 - \frac{3}{2}ha_4 + \frac{1}{6}h^2a_3 + \frac{3}{8}h^3a_2 + \frac{5}{24}h^4a_1 + \frac{19}{240}h^5a_0$$

$$A_3 = 10a_5 + ha_4 - \frac{4}{3}h^2a_3 - \frac{5}{12}h^3a_2 + \frac{1}{3}h^4a_1 + \frac{151}{360}h^5a_0$$

$$A_2 = -10a_5 + ha_4 + \frac{4}{3}h^2a_3 - \frac{5}{12}h^3a_2 - \frac{1}{3}h^4a_1 + \frac{151}{360}h^5a_0$$

$$A_1 = 5a_5 - \frac{3}{2}ha_4 - \frac{1}{6}h^2a_3 + \frac{3}{8}h^3a_2 - \frac{5}{24}h^4a_1 + \frac{19}{240}h^5a_0$$

$$A_0 = -a_5 + \frac{1}{2}ha_4 - \frac{1}{6}h^2a_3 + \frac{1}{24}h^3a_2 - \frac{1}{120}h^4a_1 + \frac{1}{720}h^5a_0$$

2. Relations between block pulse coefficients in (7.22) and (7.29). The following equations give the relations between the block pulse coefficients y_{l+i} $(i = 0, 1, \ldots, n)$ in the block pulse difference equation (7.22) and $z_{j,l}$ $(j = 0, 1, \ldots, n)$ in the block pulse regression equation (7.29). Similar relations also exist between the block pulse coefficients u_{l+i} and $v_{j,l}$.

The 1st-order system:

$$z_{1,l} = y_{l+1} - y_l$$

$$z_{0,l} = \frac{h}{2}(y_{l+1} + y_l)$$

The 2nd-order system:

$$z_{2,l} = y_{l+2} - 2y_{l+1} + y_l$$

$$z_{1,l} = \frac{h}{2}(y_{l+2} - y_l)$$

$$z_{0,l} = \frac{h^2}{6}(y_{l+2} + 4y_{l+1} + y_l)$$

The 3rd-order system:

$$z_{3,l} = y_{l+3} - 3y_{l+2} + 3y_{l+1} - y_l$$

$$z_{2,l} = \frac{h}{2}(y_{l+3} - y_{l+2} - y_{l+1} + y_l)$$

$$z_{1,l} = \frac{h^2}{6}(y_{l+3} + 3y_{l+2} - 3y_{l+1} - y_l)$$

$$z_{0,l} = \frac{h^3}{24}(y_{l+3} + 11y_{l+2} + 11y_{l+1} + y_l)$$

The 4th-order system:

$$z_{4,l} = y_{l+4} - 4y_{l+3} + 6y_{l+2} - 4y_{l+1} + y_l$$

$$z_{3,l} = \frac{h}{2}(y_{l+4} - 2y_{l+3} + 2y_{l+1} - y_l)$$

$$z_{2,l} = \frac{h^2}{6}(y_{l+4} + 2y_{l+3} - 6y_{l+2} + 2y_{l+1} + y_l)$$

$$z_{1,l} = \frac{h^3}{24}(y_{l+4} + 10y_{l+3} - 10y_{l+1} - y_l)$$

$$z_{0,l} = \frac{h^4}{120}(y_{l+4} + 26y_{l+3} + 66y_{l+2} + 26y_{l+1} + y_l)$$

The 5th-order system:

$$z_{5,l} = y_{l+5} - 5y_{l+4} + 10y_{l+3} - 10y_{l+2} + 5y_{l+1} - y_l$$

$$z_{4,l} = \frac{h}{2}(y_{l+5} - 3y_{l+4} + 2y_{l+3} + 2y_{l+2} - 3y_{l+1} + y_l)$$

$$z_{3,l} = \frac{h^2}{6}(y_{l+5} + y_{l+4} - 8y_{l+3} + 8y_{l+2} - y_{l+1} - y_l)$$

$$z_{2,l} = \frac{h^3}{24}(y_{l+5} + 9y_{l+4} - 10y_{l+3} - 10y_{l+2} + 9y_{l+1} + y_l)$$

$$z_{1,l} = \frac{h^4}{120}(y_{l+5} + 25y_{l+4} + 40y_{l+3} - 40y_{l+2} - 25y_{l+1} - y_l)$$

$$z_{0,l} = \frac{h^5}{720}(y_{l+5} + 57y_{l+4} + 302y_{l+3} + 302y_{l+2} + 57y_{l+1} + y_l)$$

References

Atkinson, K.E. (1985): *Elementary numerical analysis,* John Wiley & Sons, New York.

Boxer, R. and Thaler, S. (1956): *A simplified method of solving linear and nonlinear systems,* Proc. Inst. Radio Engrs, vol.44, pp.89-101.

Bryson, A.E. and Ho, Y.-C. (1975): *Applied Optimal Control,* John Wiley & Sons, New York.

Chen, C.F. and Hsiao, C.H. (1975): *Walsh series analysis in optimal control,* Int. J. Control, vol.21, pp.881-897.

Chen, C.F., Tsay, Y.T. and Wu, T.T. (1977): *Walsh operational matrices for fractional calculus and their application to distributed systems,* J. Franklin Inst., vol.303, pp.267-284.

Chen, W.-L. (1983): *Analysis of the delay systems without time partition,* J. Chinese Inst. Engrs, vol.6, pp.61-64.

Chen, W.-L. and Chung, C.-Y. (1987): *New integral operational matrix in block pulse series analysis,* Int. J. Systems Sci., vol.18, pp.403-408.

Chen, W.-L. and Hsu, C.-S. (1987): *Convergence of the block pulse series solution of a linear distributed parameter system,* Int. J. Systems Sci., vol.18, pp.965-975.

Chen, W.-L. and Jeng, B.-S. (1981): *Analysis of piecewise constant delay systems via block pulse functions,* Int. J. Systems Sci., vol.12, pp.625-633.

Chen, W.-L. and Lin, J.-F. (1986): *Analysis and identification of systems with a nonlinear element,* Int. J. Systems Sci., vol.17, pp.1097-1104.

Chen, W.-L. and Meng, C.-H. (1982): *A general procedure of solving the linear delay system via block pulse functions,* Comput. and Elect. Engng, vol.9, pp.153-166.

Chen, W.-L. and Wu, S.-G. (1985): *Analysis of multirate sampled-data systems by block pulse functions,* Mathematics and Computers in Simulation, vol.27, pp.503-510.

Cheng, B. and Hsu, N.-S. (1982a): *Analysis and parameter estimation of bilinear systems via block pulse functions,* Int. J. Control, vol.36, pp.53-65.

Cheng, B. and Hsu, N.-S. (1982b): *Single-input single-output system identification via block pulse functions,* Int. J. Systems Sci., vol.13, pp.697-702.

Curtain, R.F. and Pritchard, A.J. (1977): *Functional analysis in modern applied mathematics,* Academic Press, London.

Dalton O.N. (1978): *Further comments on "Design of piecewise constant gains for optimal control via Walsh functions",* IEEE Trans. on Automatic Control, vol.23, pp.760-762.

Frank, P.M. (1978): *Introduction to system sensitivity theory,* New York: Academic Press.

Franklin, G.F., Powell, J.D. and Workman, M.L. (1990): *Digital control of dynamic systems,* second edition, Addison-Wesley, Massachusetts.

Gopalsami, N. and Deekshatulu, B.L. (1976a): *Comments on "Design of piecewise constant gains for optimal control via Walsh functions",* IEEE Trans. on Automatic Control, vol.21, pp.634-636.

Gopalsami, N. and Deekshatulu, B.L. (1976b): *Time-domain synthesis via Walsh functions,* IEE Proceedings, vol.123, pp.461-462.

Halijak, C.A. (1960): *Digital approximation of solutions of differential equations using trapezoidal convolution,* Report IT-64, Bendix Systems Div., Ann Arbor, Mich.

Harmuth, P. (1969): *Application of Walsh functions in communication,* IEEE Spectrum, vol.6, pp.82-91.

Haykin, S.S. (1972): *A unified treatment of recursive digital filtering,* IEEE Trans. on Automatic Control, vol.17, pp.113-116.

Hildebrand, F.B. (1974): *Introduction to numerical analysis,* second edition, McGraw-Hill, New York.

Hsia, T.C. (1968): *Identification of parameters in linear systems with transport lags,* Proc. 11th Midwest Symposium on Circuit Theory, May 13-14, Notre Dame, U.S.A., pp.62-67.

Hsu, N.-S. and Cheng, B. (1981): *Analysis and optimal control of time-varying linear systems via block pulse functions,* Int. J. Control, vol.33, pp.1107-1122.

Hsu, N.-S. and Cheng, B. (1982): *Identification of non-linear distributed systems via block pulse functions,* Int. J. Control, vol.36, pp.281-291.

Hwang, C. and Guo, T.-Y. (1984a): *Identification of lumped linear time-varying systems via block pulse functions,* Int. J. Control, vol.40, pp.571-583.

Hwang, C. and Guo, T.-Y. (1984b): *New approach to the solution of integral equations via block pulse functions,* Int. J. Systems Sci., vol.15, pp.361-373.

Hwang, C. and Shih, Y.-P. (1986): *On the operational matrices of block pulse functions,* Int. J. Systems Sci., vol.17, pp.1489-1498.

Hwang, C. and Shih, Y.P. (1985): *Optimal control of delay systems via block pulse functions,* J. Optimization Theory and Applications, vol.45, pp.101-112.

Hwang, C., Guo, T.-Y. and Shih, Y.-P. (1983): *Numerical inversion of multidimensional Laplace transforms via block pulse functions,* IEE Proceedings, vol.130, pp.250-254.

Hwang, C., Shih, D.-H. and Kung, F.-C. (1986): *Use of block pulse functions in the optimal control of deterministic systems,* Int. J. Control, vol.44, pp.343-349.

Jan, T.G. and Wong, K.M. (1981): *Bilinear system identification by block pulse functions,* J. Franklin Inst., vol.312, pp.349-359.

Jaw, Y.-G. and Kung, F.-C. (1984): *Identification of a single-variable linear time-varying system via block pulse functions,* Int. J. Systems Sci., vol.15, pp.885-893.

Jiang, Z.H. (1986): *Use of block pulse functions for output sensitivity analysis of linear systems,* Int. J. Control, vol.44, pp.407-417.

Jiang, Z.H. (1987a): *Block pulse function approach to the identification of MIMO-systems and time-delay systems,* Int. J. Systems Sci., vol.18, pp.1711-1720.

Jiang, Z.H. (1987b): *Identification of continuous non-linear systems via block pulse functions,* 30th Midwest Symposium on Circuits and Systems, August 16-18, Syracuse, U.S.A., pp.36-39.

Jiang, Z.H. (1987c): *New approximation method for inverse Laplace transforms using block pulse functions,* Int. J. Systems Sci., vol.18, pp.1873-1888.

Jiang, Z.H. (1988): *Block pulse function approach for the identification of Hammerstein model non-linear systems,* Int. J. Systems Sci., vol.19, pp.2427-2439.

Jiang, Z.H. (1990): *System identification via block pulse difference equations,* Ph.D. thesis, Swiss Federal Institute of Technology Zurich.

Jiang, Z.H. and Schaufelberger, W. (1985a): *A new algorithm for single-input single-output system identification via block pulse functions,* Int. J. Systems Sci., vol.12, pp.1559-1571.

Jiang, Z.H. and Schaufelberger, W. (1985b): *Recursive formula for the multiple integral using block pulse functions,* Int. J. Control, vol.41, pp.271-279.

Jiang, Z.H. and Schaufelberger, W. (1985c): *Design of adaptive regulators based on block pulse function identification,* IEE International Conference "Contral 85", July 9-11, Cambridge, England, pp.581-586.

Jiang, Z.H. and Schaufelberger, W. (1991a): *Recursive block pulse function method,* in "Identification of continuous-time systems", Editor: Sinha, N.K. and Rao G.P., Kluwer Academic Publishers, pp.205-225.

Jiang, Z.H. and Schaufelberger, W. (1991b): *Recursive computational algorithms for a set of block pulse operational matrices,* Technical Report, IDA Project Center, ETH Zurich.

Jiang, Z.H. and Schaufelberger, W. (1991c): *A recursive identification method for continuous time-varying linear systems,* Proc. 34th Midwest Symposium on Circuits and Systems, May 14-17, Monterey, U.S.A.

Kawaji, S. (1983): *Block pulse series analysis of linear systems incorporating observers,* Int. J. Control, vol.37, pp.1113-1120.

Kekkeris, G.T. and Marszalek, W. (1989): *Sensitivity analysis of systems involving λ-variation via block pulse functions,* Int. J. Systems Sci., vol.20, pp.2003-2009.

Korevaar, J. (1968): *Mathematical methods,* vol.1, Academic Press, New York.

Kraus, F. and Schaufelberger, W. (1990): *Identification with block pulse functions, modulating functions and differential operators,* Int. J. Control, vol.51, pp.931-942.

Kung, F.-C. and Shih, D.-H. (1986): *Analysis and identification of Hammerstein model non-linear delay systems using block pulse function expansions,* Int. J. Control, vol.43, pp.139-147.

Kung, F.C. and Chen, S.Y. (1978): *Solution of integral equations using a set of block pulse functions,* J. Franklin Inst., vol.306, pp.283-291.

Kwong, C.P. and Chen, C.F. (1981a): *The convergence properties of block pulse series,* Int. J. Systems Sci., vol.12, pp.745-751.

Kwong, C.P. and Chen, C.F. (1981b): *Linear feedback system identification via block pulse functions,* Int. J. Systems Sci., vol.12, pp.635-642.

Lighthill, M.J. (1958): *Introduction to Fourier analysis and generalized functions,* Cambridge University Press.

Liou, C.-T. and Chou, Y.-S. (1988): *Successive parameter estimation of continuous dynamic systems,* Int. J. Systems Sci., vol.19, pp.1149-1158.

Madwed, A. (1950): *Number series method of solving linear and nonlinear differential equations,* Report 6445-T-26, Instrumentation Lab., MIT, Cambridge, Massachusetts.

Marszalek, W. (1983): *The block pulse functions method of the two-dimensional Laplace transform,* Int. J. Systems Sci., vol.14, pp.1311-1317.

Marszalek, W. (1984a): *Block pulse functions method of the inverse Laplace transform for irrational and transcendental transfer functions,* J. Franklin Inst., vol.318, pp.193-200.

Marszalek, W. (1984b): *On the inverse Laplace transform of irrational and transcendental transfer functions via block pulse functions method,* Int. J. Systems Sci., vol.15, pp.869-876.

Marszalek, W. (1984c): *On the nature of block pulse operational matrices,* Int. J. Systems Sci., vol.15, pp.983-989.

Marszalek, W. (1985a): *Analysis of bilinear systems with Picard's method and block pulse operational matrices,* J. Franklin Inst., vol.320, pp.105-109.

Marszalek, W. (1985b): *On the nature of block pulse operational matrices: some futher results*, Int. J. Systems Sci., vol.16, pp.727-743.

Nachbin, L. (1981): *Introduction to functional analysis: Banach spaces and differential calculus*, Marcel Dekker, Inc., New York.

Nath, A.K. and Lee, T.T. (1983): *On the multidimensional extension of block pulse functions and their completeness*, Int. J. Systems Sci., vol.14, pp.201-208.

Ning, J. and Jiong, J. (1988): *Parameter or non-parameter identification of distributed parameter systems via block pulse functions*, Int. J. Systems Sci., vol.19, pp.1039-1045.

Oldham, K.B. and Spanier, J. (1974): *The fractional calculus*, Academic Press, Orlando.

Pagaspe, J.P. (1976): *Contributions á l'identification non-lineaire per la représentation fonctionnelle et la transformée de Laguerre*, Ph.D. thesis, Bordeaux-I University.

Palanisamy, K.R. (1983): *A note on block pulse function operational matrix for integration*, Int. J. Systems Sci., vol.14, pp.1287-1290.

Palanisamy, K.R. and Arunachalam, V.P. (1985): *Solution of variational problems using block pulse functions*, Int. J. Systems Sci., vol.16, pp.257-267.

Palanisamy, K.R. and Bhattacharya, D.K. (1981): *System identification via block pulse functions*, Int. J. Systems Sci., vol.12, pp.643-647.

Ren, F., Shih, Y.-P., Pei, S.-C. and Guo, R.-T. (1983): *Parameter Identification of bilinear systems by block pulse functions*, J. Chinese Inst. Engrs, vol.6, pp.39-42.

Perng, M.-H. and Chen, W.-L. (1985): *Block pulse series solution of simultaneous first-order partial differential equations*, Int. J. Systems Sci., vol.16, pp.1573-1580.

Rao, B.R. and Ganapathy, S. (1979): *Linear time-varying systems: state transition matrix*, IEE Proceedings, vol.126, pp.1331-1335.

Rao, G.P. (1983): *Piecewise constant orthogonal functions and their application to systems and control*, Springer-Verlag, Berlin.

Rao, G.P. and Srinivasan, T. (1978a): *Analysis and synthesis of dynamic systems containing time delays via block pulse functions*, IEE Proceedings, vol.125, pp.1064-1068.

Rao, G.P. and Srinivasan, T. (1978b): *Remarks on "Author's reply" to "Comments on 'Design of piecewise constant gains for optimal control via Walsh functions' "*, IEEE Trans. on Automatic Control, vol.23, pp.762-763.

Rao, G.P. and Srinivasan, T. (1980a): *An optimal method of solving differential equations characterizing the dynamics of a current colletion system for an electric locomotive*, J. Inst. Maths. Applics., vol.25, pp.329-342.

Rao, G.P. and Srinivasan, T. (1980b): *Multidimensional block pulse functions and their use in the study of distributed parameter systems*, Int. J. Systems Sci., vol.11, pp.689-709.

Rao, G.P. (1978): *Solution of certain nonlinear functional equations via block pulse functions,* Proc. 5th National Systems Conference, India, pp.287-290.

Rao, V.P. and Rao, K.R. (1979): *Optimal feedback control via block pulse functions,* IEEE Trans. on Automatic Control, vol.24, pp.372-374.

Rosko, J.S. (1972): *Digital simulation of physical systems,* Addison-Wesley, New York.

Sannuti, P. (1977): *Analysis and synthesis of dynamic systems via block pulse functions,* IEE Proceedings, vol.124, pp.569-571.

Shieh, L.S. and Yates, R.E. (1979): *Solving inverse Laplace transform, linear and non-linear state equations using block pulse functions,* Comput. and Elect. Engng, vol.6, pp.3-17.

Shieh, L.S., Schneider, W.P. and Williams, D.R. (1971): *A chain of factored matrices for Routh array inversion and continued fraction inversion,* Int. J. Control, vol.13, pp.691-703.

Shieh, L.S., Yates, R.E. and Navarro, J.M. (1978): *Representation of continuous-time state equations by discrete-time state equations,* IEEE Trans. on Systems, Man and Cybernetics, vol.8, pp.485-492.

Shieh, L.S., Yeung, C.K. and McInnis, B.C. (1978): *Solution of state-space equations via block pulse functions,* Int. J. Control, vol.28, pp.383-392.

Shih, Y.-M. (1978): *Block pulse function analysis of time-varying and non-linear networks,* J. Chinese Inst. Engrs, vol.1, pp.43-52.

Shih, Y.-P. (1981): *Optimal control of time-varying linear systems via block pulse functions,* Proc. Natl. Sci. Counc., vol.5, pp.130-136.

Shih, Y.-P. and Hwang, C. (1982): *Application of block pulse functions in dynamic simulation,* Computers and Chemical Engineering, vol.6, pp.7-13.

Shih, Y.-P., Hwang, C. and Chia, W.-K. (1980): *Parameter estimation of delay systems via block pulse functions,* J. Dynamic Systems, Measurement and Control, vol.102, pp.159-162.

Shilov, G.E. and Gurevich, B.L. (1966): *Integral, measure and derivative: A unified approach,* Prentice Hall, Englewood Cliffs.

Sinha, N.K. and Zhou, Q.-J. (1983): *Discrete-time approximation of multivariable continuous-time systems,* IEE Proceedings, vol.130, pp.103-110.

Sinha, N.K. and Zhou, Q.-J. (1984): *State estimation using block pulse functions,* Int. J. Systems Sci., vol.15, pp.341-350.

Söderström, T. and Stoica, P.G. (1983): *Instrumental variable methods for system identification,* Springer-Verlag, Berlin.

Stavroulakis, P. and Tzafestas, S (1980): *Distributed-parameter observer-based control implementation using finite spatial measurements,* Math. Computers Simulation, vol.22, pp.373-379.

Sun, Y.Y. (1981): *Solution of convolution integrals and correlation functions via block pulse functions*, J. Chinese Inst. Engrs, vol.4, pp.31-38.

Titchmarsh, E.C. (1960): *The theory of functions*, Oxford University Press.

Tolstov, G.P. (1962): *Fourier series*, Prentice Hall, Englewood Cliffs.

Tuma, J.J. (1987): *Engineering mathematics handbook*, McGraw-Hill Book Company, New York.

Tustin, A. (1947): *A method of analysing the behavior of linear systems in terms of time series*, J. Inst. Elect. Engrs, vol.94(IIA), pp.130-142.

Tzafestas, S.G., Papastergiou, C.N. and Anoussis, J.N. (1984): *Dynamic reactivity computation in nuclear reactors using block pulse function expansion*, Int. J. Modelling and Simulation, vol.4, pp.73-76.

Wang, C.-H. (1982): *Generalized block pulse operational matrices and their applications to operational calculus*, Int. J. Control, vol.36, pp.67-76.

Wang, C.-H. (1983): *On the generalization of block pulse operational matrices for fractional and operational calculus*, J. Franklin Inst., vol.315, pp.91-102.

Wang, C.-H. and Marleau, R.S. (1985): *System identification via generalized block pulse operational matrices*, Int. J. Systems Sci., vol.16, pp.1425-1430.

Wang, C.-H. and Marleau, R.S. (1987): *Recursive computational algorithm for the generalized block pulse operational matrix*, Int. J. Control, vol.45, pp.195-201.

Wang, C.-H. and Shih, Y.-P. (1982): *Explicit solutions of integral equations via block pulse functions*, Int. J. Systems Sci., vol.13, pp.773-782.

Wang, S.Y. (1982a): *The application of block pulse function in identification of nonlinear time varying systems*, American Control Conference, pp.439-440.

Wang, S.Y. (1982b): *The application of block pulse function in identification of parameters of nonlinear time varying systems*, Proc. of 6th IFAC Symposium on Identification and System Parameter Estimation, Washington D.C., USA, pp.1045-1050.

Wang, S.Y. (1983a): *Block pulse operator and its application in control theory (I)*, J. of East China Institute of Chemical Technology, pp.1-15.

Wang, S.Y. (1983b): *Block pulse operator and its application in control theory (II)*, J. of East China Institute of Chemical Technology, pp.501-509.

Wang, S.Y. (1990): *Convergence of block pulse series approximation solution for optimal control problem*, Int. J. Systems Sci., vol.21, pp.1355-1368.

Wang, S.Y. (1991): *Use of the block pulse operator*, in "Identification of continuous-time systems", Editor: Sinha, N.K. and Rao G.P., Kluwer Academic Publishers, pp.159-203.

Wang, S.Y. and Jiang, W.S. (1984): *The application of block pulse operator in identification of distributed parameter systems,* Proc. of 9th IFAC Triennial World Congress, Budapest, Hungary, pp.655-660.

Wang, S.Y. and Jiang, W.S. (1985a): *Identification of nonlinear distributed parameter systems using block pulse operator,* Proc. of 7th IFAC Symposium on Identification and System Parameter Estimation, York, U.K., pp.803-808.

Wang, S.Y. and Jiang, W.S. (1985b): *Identification and optimal control of penicillin fermentation process – an application of block pulse operator,* Proc. of Int. Conference on Industry Process Modelling and Control, vol.II, pp.163-170.

Wouters, W.R. (1972): *On-line identification in an unknown stochastic environment,* IEEE Trans. on Systems,Man and Cybernetics, vol.2, pp.666-668.

Wu, M.-Y. and Sherif, A. (1976): *On the commutative class of linear time-varying systems,* Int. J. Control, vol.23, pp.433-444.

Wu, W.-T. and Juang, L.-Y. (1980): *On recursive parameter estimation of multivariable systems,* J. Chinese Inst. Engrs, vol.3, pp.89-93.

Wu, W.-T. and Wong, Y.-S. (1980): *Recursive parameter estimation of time-lag systems via block pulse function,* Computers and Chemical Engineering, vol.4, pp.201-203.

Xing, J.-X. and Wang, X.-T. (1985): *Analysis of nonlinear systems and synthesis of optimal control via block pulse functions,* ACTA Automatica Sinica, vol.11, pp.175-183.

Xu, N.-S. and Zheng, B. (1982): *Analysis and optimal control of time-varying linear systems using block pulse functions,* ACTA Automatica Sinica, vol.8, pp.55-67.

Zeng, G.-D. (1981): *Representation of image scanning function and calculation of convolution integral by block pulse functions,* J. of Huazhong University of Science and Technology, vol.9, pp.27-34.

Zeng, G.-D. and Zhang, Y.-L. (1985a): *A class of transfer matrices for linear systems and the recursive calculation,* J. of Huazhong University of Science and Technology, vol.13, pp.23-30.

Zeng, G.-D. and Zhang, Y.-L. (1985b): *A new kind of block diagram of linear systems with operation,* Proc. of China 1985 International Conference on Circuits and Systems, Beijing, China, pp.435-439.

Zeng, G.-D. and Zhang, Y.-L. (1985c): *A novel mathematical model and block diagram,* J. of Huazhong University of Science and Technology, vol.13, pp.31-38.

Zhang, H.Y. and Chen, J. (1990): *Identification of physical parameters of ground vehicle using block pulse function method,* Int. J. Systems Sci., vol.21, pp.631-642.

Zhou, Q.-J. and Sinha, N.K. (1982): *Identification of continuous-time state-space model from samples of input-output data,* Electronics Letters, vol.18, pp.50-51.

Zhu, J.-M. and Lu, Y.-Z. (1987): *New approach to hierarchical control via block pulse transformation,* Int. J. Control, vol.46, pp.441-453.

Zhu, J.-M. and Lu, Y.-Z. (1988b): *Hierarchical optimal control for distributed parameter systems via block pulse operator,* Int. J. Control, vol.48, pp.685-703.

Index

Lecture Notes in Control and Information Sciences

Edited by M. Thoma and A. Wyner

Lecture Notes in Control and Information Sciences

Edited by M. Thoma and A. Wyner

Lecture Notes in Control and Information Sciences

Edited by M. Thoma and A. Wyner

Vol. 174: A.J.M. Beulens, H.-J. Sebastian (Eds.)
Optimization-Based Computer-Aided
Modelling and Design
Proceedings of the First Working Conference
of the IFIP TC 7.6 Working Group,
The Hague, The Netherlands, 1991
VIII, 270 pages, 1992

Vol. 175: E. Rogers, D.H. Owens
Stability Analysis for Linear Repetitive Processes
VII, 197 pages, 1992

Vol. 176: B.L. Rozovskii, R.B. Sowers (Eds.)
Stochastic Partial Differential Equations
and Their Applications
Proceedings of IFIP WG 7/1 International Conference
University of North Carolina at Charlotte, NC
June 6 - 8, 1991
VIII, 251 pages, 1992

Vol. 177: I. Karatzas, D. Ocone (Eds.)
Applied Stochastic Analysis
Proceedings of a US-French Workshop,
Rutgers University, New Brunswick, N.J.
April 29 - May 2, 1991
X, 311 pages, 1992

Vol. 178: J.P. Zolésio (Ed.)
Boundary Control and Boundary Variation
Proceedings of IFIP WG 7.2 Conference,
Sophia Antipolis, France
October 15 - 17, 1990
VIII, 392 pages 1992

Vol. 179: Z.H. Jiang, W. Schaufelberger
Block Pulse Functions and Their Applications
in Control Systems
XII, 237 pages, 1992

Lecture Notes in Control and Information Sciences